African Women, ICT and Neoliberal Politics

How can we promote people-centered governance in Africa? Cell phones/information and communications technology (ICT) are shown to be linked to neoliberal understandings of more democratic governance structures, defined by the Worldwide Governance Indicators as: the rule of law, corruption-control, regulation quality, government effectiveness, political stability/no violence, and voice and accountability. However, these indicators fall short: they do not emphasize gender equity or pro-poor policies.

Writing from an African feminist scholar-activist perspective, Assata Zerai emphasizes the voices of women in two ways: (1) she examies how *women's* access to ICT makes a difference to the success of people-centered governance structures; and (2) she demonstrates how African women's scholarship, too often marginalized, must be used to expand and redefine the goals and indicators of democratic governance in African countries.

Challenging the status quo that praises the contributions of cell phones to the diffusion of knowledge and resultant better governance in Africa, this book is an important read for scholars of politics and technology, gender and politics, and African Studies.

Assata Zerai is Professor of Sociology at the University of Illinois at Urbana-Champaign. Her interests have included maternal and child health (MCH), health activism, safe water and sanitation, ICT in Africa and the African Diaspora, and making the intellectual work of African women scholars and activists more accessible; as well as U.S.-based studies of MCH, Black feminist praxis, and diversity and LGBTIQ inclusiveness in Protestant congregations. Her recent books include *Safe Water, Sanitation and Early Childhood Malnutrition in East Africa: An Africana Feminist Analysis of the lives of Women and Children in Kenya, Tanzania and Uganda* (Zerai and Brenda N. Sanya, eds, Rowman & Littlefield, Lexington Books, 2018); *Intersectionality in Intentional Communities: The Struggle for Inclusivity in Multicultural U.S. Protestant Congregations* (Rowman & Littlefield, Lexington Books, 2016); and *Hypermasculinity and State Violence in Zimbabwe: An Africana Feminist Analysis of Maternal and Child Health* (Africa World Press, 2014).

Routledge Studies on Gender and Sexuality in Africa

1. **The Tunisian Women's Rights Movement**
   From Nascent Activism to Influential Power-broking
   Jane D. Tchaïcha and Khédija Arfaoui

2. **Disability and Sexuality in Zimbabwe**
   Voices from the Periphery
   Christine Peta

3. **Love, Sex and Teenage Sexual Cultures in South Africa**
   16 Turning 17
   Deevia Bhana

4. **African Women, ICT and Neoliberal Politics**
   The Challenge of Gendered Digital Divides to
   People-Centered Governance
   Assata Zerai

African Women, ICT and Neoliberal Politics

The Challenge of Gendered Digital Divides to People-Centered Governance

Assata Zerai

Routledge
Taylor & Francis Group
LONDON AND NEW YORK

First published 2019
by Routledge
2 Park Square, Milton Park, Abingdon, Oxon OX14 4RN

and by Routledge
711 Third Avenue, New York, NY 10017

Routledge is an imprint of the Taylor & Francis Group, an informa business

British Library Cataloguing-in-Publication Data
A catalogue record for this book is available from the British Library

Library of Congress Cataloging-in-Publication Data
Names: Zerai, Assata, 1964– author.
Title: African women, ICT and neoliberal politics : the challenge of gendered digital divides to people-centered governance / Assata Zerai.
Other titles: Routledge studies on gender and sexuality in Africa ; 4.
Description: New York City : Routledge, 2018. |
Series: Routledge studies on gender and sexuality in Africa ; 4 |
Includes bibliographical references.
Identifiers: LCCN 2018019006 | ISBN 9781138559363 (hardback) |
ISBN 9780203712856 (ebook) | ISBN 9781351363648 (mobipocket)
Subjects: LCSH: Women–Africa–Social conditions. |
Women's rights–Africa. | Sex discrimination–Africa. |
Telecommunication–Africa. | Women–Political activity–Africa.
Classification: LCC HQ1787 .Z47 2018 | DDC 305.42096–dc23
LC record available at https://lccn.loc.gov/2018019006

ISBN: 978-1-138-55936-3 (hbk)
ISBN: 978-0-203-71285-6 (ebk)

Typeset in Times New Roman
by Out of House Publishing

Contents

Figures

Tables

Acronyms

AfDB	African Development Bank Group
CEDAW	Convention on the Elimination of All Forms of Discrimination against Women
CPIA	Country Policy and Institutional Assessment Index
DHS	Demographic and Health Surveys
ICT	Information and Communications Technology
IDHS	Integrated Demographic and Health Surveys
IPV	Intimate partner violence
MWDHS	Malawi Demographic and Health Surveys
TDHS	Tanzania Demographic and Health Surveys
WDI	World Development Indicators
WGI	Worldwide Governance Indicators
ZDHS	Zimbabwe Demographic and Health Surveys

Foreword

How can we promote people-centered governance on the Africa continent and beyond? Written from an African feminist scholar-activist perspective, this work emphasizes the voices of women and demonstrates how their framings are important when we consider the impact of science and technology in society. A mechanism that impedes the influence of African feminist thought and decelerates its impact on governance is examined, namely that encumbered access to information and communications technology (ICT) among women on the African continent (and in its Diaspora)[1] creates barriers to communal discourse and problem solving.

Asongu and Nwachukwu (2016) have recently shown that cell phones/ICT are linked to neoliberal understandings of more democratic governance structures, defined Worldwide Governance Indicators (WGI) and promoted by the World Bank as: the rule of law, corruption-control, regulation quality, government effectiveness, political stability/no violence and voice and accountability by (Kaufmann et al. 2010). However, this definition deviates from more social justice oriented governance approaches. I argue that in key areas these androcentric neoliberal indicators of democratic governance fall short; they do not emphasize gender equity or pro-poor policies seen in social justice approaches to democratic governance. To better understand and promote people-centered governance in Africa, in this book I emphasize the voices of women in two ways: (1) I examine how women's access to ICT makes a difference to the success of creating more people-centered governance structures; and (2) I demonstrate how African women's and gender-sensitive scholarship, too often marginalized, must be used to expand and redefine the goals and indicators of democratic governance in African countries.

I examine whether ICT is attached to governance structures that serve the most marginalized members of society, and especially women.

Are they attached to collective aspects and individual aspects of society that are most important to women? For example, pro-poor policies and attention to disability services would be a part of a governance structure that serves the most vulnerable. Amina Mama (1998) talks about those issues that are most important to women. Some include marriage and safety in marriage, socioeconomic status, children's wellbeing, and women's health. Further, a focus on addressing the needs of single mothers would also be considered. Another question to address is the types of information and even continuing education that people can access with ICT in African societies, especially information and education that is most relevant to women at the bottom of the economic pyramid. Above all, it is important to understand local definitions of good governance.

This work promises to contribute importantly to the social scientific literature because it builds from the vibrant intellectual work of African feminists to guide quantitative analysis. It enlarges the scope of empirical and theoretically grounded studies within the field of science and technology in society by addressing ICT, improved governance, and impacts on women's lives in understudied regions of the world. Further, this work elicits results that give rise to useful development effectiveness and ICT-related policy recommendations.

Challenging the status quo that praises the contributions of cell phones to the diffusion of knowledge and resulting better governance in Africa, this book is an important read for scholars of politics and technology, gender and politics, and African Studies.

Note

1 As is the case for many women on the continent, women in the African Diaspora with inadequate resources also live in what Olu Oguibe (2003) describes as the digital Third World characterized by exclusion and nonconnectivity, despite their location in industrialized countries.

Introduction

In the traditions of Professor Victor C. Uchendu and Professor Ifi Amadiume: African women and the challenge of digital divides to people-centered governance[1]

I had the distinct honor to participate in a celebration to commemorate the fiftieth anniversary of the release of the late Professor Victor Uchendu's *The Igbo of Southeastern Nigeria.* Professor Uchendu has left a very long and deep legacy in the fields of anthropology, sociology, and, of course African Studies. Professor Mbye Cham, who was chair of the Department of African Studies at Howard University, the oldest African Studies doctoral program in the U.S., indicates that Professor Uchendu recruited him into African Studies at the University of Illinois, when Professor Uchendu chaired the department there. Of course, symposium organizer Professor Biko Agozino was Professor Uchendu's student. And I too followed in Prof's footsteps, as the former director of the Center for African Studies at the University of Illinois.

In the field of ethnography, one in which Professor Uchendu was a pioneer, we now speak of emic and etic perspectives. In 1967, linguist Kenneth Pike wrote "Etic viewpoint studies behavior as from outside of a particular system," while the "emic viewpoint results from studying behavior as from inside the system." Professor Uchendu wrote from an emic or insider's perspective. He indicated:

> In the following chapters, the readers will be introduced to Igbo society and culture by one of its culture-bearers who has been privileged to gather his data as a "full participant" and tries to write as a "man of science."
>
> (Uchendu 1965: 4)

Some scholars challenged the value of Uchendu gathering data as a "full participant." The veracity of the data and the objectivity of the analyst were questioned by positivists.

Positivists purport that social science must be "scientific" in the sense as in the physical and experimental sciences. And they also claim that research must be value-free. Harding and Norberg point out that given our limited points of view, value-free research is an unachievable ideal:

> Dominant groups are especially poorly equipped to identify oppressive features of their own beliefs and practices, as standpoint methodologists have argued. Their activities in daily life do not provide them with the intellectual and political resources necessary to detect such values and interests in their own work.
>
> (Harding and Norberg 2005: 2)

Value-free research is *not only* an unachievable ideal. Feminists maintain that it is also an *undesirable* one. In fact, socially engaged research, i.e., research ethically and politically accountable for its consequences, *can* produce knowledge (Agozino 1999; Harding and Norberg 2005). Professor Uchendu was a part of the *postpositive movement* and held himself accountable to African audiences and disadvantaged groups. In Table 0.1, I have repeated and updated a distinction Rae Banks and I have drawn between positivist and interpretive/postpositivist approaches in earlier work (Zerai and Banks 2002).

Sociologist Max Weber is known for establishing antipositive traditions (of which postpositivism is now a part) in the social sciences. But of course, we know that W.E.B. Du Bois wrote at length about

Table 0.1 Epistemological differences between positivist and interpretive approaches

Positivism	*Interpretive, postpositivism*
Researcher distanced from the object of study	**Socially engaged research** (Harding and Norberg 2005) **Committed objectivity** (Agozino 1999): an ethic of caring and personal accountability (Collins 1990)
Absence of emotions	**Lived experience** as criteria for meaning (Collins 1990; Uttal and Cuádraz 1999)—emic perspective is valued (Uchendu 1965; Amadiume 1987, 1997, 2000; Zerai 2000a, 2000b)
Values largely withdrawn from the research process ("value-free")	**Marginalized voices/perspectives** are agents of knowledge (Collins 1990, 1998a, 1998b; Mama 2002), including double consciousness (Du Bois 1903b)
Adversarial debates to ascertain truth	**Humanizing speech** to assess knowledge claims (hooks and West 1991)

double consciousness, the quintessential representation of not only the conflict between insider and outsider perspectives, but also the African and African Diaspora scholar as an embodiment of both perspectives. Now, with interpretive approaches, many have dispensed with the ideal of a value-free social science. Due to the work of Du Bois, Uchendu, Harding, and others, insiders' perspectives are now valued within some sociological traditions and among growing numbers of contemporary social scientists. African and African Diaspora scholars have had to work hard to have our work accepted by the wider social sciences establishment. Professor Uchendu's *The Igbo of Southeastern Nigeria* is written in the tradition of the work of Cheikh Anta Diop, who wrote three dissertations until one was finally accepted (Clarke 1974, 1990; van Sertima 1989) as a part of African and African American traditions in African Studies of scholars who would/will not be deterred by Western devaluing of Africa and African thought. Professor Uchendu's early persistence is described by Biko Agozino (2007: 290).

> He [Professor Uchendu] exhibited this radical independence of spirit when he wanted to write his dissertation on his beloved Igbo culture, but his supervisors at Northwestern University insisted that he should do fieldwork among Native Americans because his cultural baggage would not allow him to be completely objective while studying his own Igbo culture. He agreed and studied Native Americans, making contributions to comparative anthropology in the process with the discovery that the so-called cultural baggage was never abandoned in other cultures, where he found himself comparing the inquisitiveness of the Igbo with the aloofness of Native Americans in relation to strangers. Remarkably, he completed his book about the Igbo first within a year and it became the instant classic that made his name, earning him the honor of being appointed as the founding Director of the famous Center for African Studies at the University of Illinois at Urbana-Champaign.
> (Agozino 2007: 290)

Professor Uchendu's work was prophetic as it focused on women's agency among the Igbo in his 1965 publication, and he continued in this vein in his subsequent works that highlight women's contributions, including "Ezi Na Ulo: The Extended Family in Igbo Civilization" (2007). He inspired many who came after him, including Professor Ifi Amadiume (Obbo 1999), who followed in this tradition in the publication of her works such as the classic *Male Daughters, Female Husbands: Gender and Sex in African Society* (1987); *Re-inventing*

African Matriarchy, Religion, and Culture (1997); and *Daughters of the Goddess, Daughters of Imperialism: African Women Struggle for Culture, Power and Democracy* (2000). Both Uchendu and Amadiume wrote from an emic perspective about their beloved Nigeria/West Africa. While my work is not largely qualitative, I too, like Uchendu, Diop, Du Bois, Anna J. Cooper[2], Amadiume and others, seek to write as a "full participant," as a citizen of the pan-African world who takes the issues of women and development very personally, and as a "woman of science." In other work, I have written at length about my scholar-activist orientation to my intellectual work. I too have endured academic criticism for my *committed* objectivity (as described by Agozino). And just like Uchendu and others, I have persisted despite many barriers. Following the legacies of Anna J. Cooper, Audre Lorde, Professor Uchendu, and contemporary scholars such as Professor Ifi Amadiume, this book focuses on the need to center the concerns of African women, and for African women activists' and scholars' voices to come to the forc.

African feminism

African and Africana feminism[3] (Gaidzanwa 1997; Mama 2002; Zerai 2014) articulate women-centered perspectives that describe intersecting oppressions and privileges of African women's and girls' lived experiences as well as the social structures in which they are situated. In prior publications, I (Zerai 2014 and Zerai et al. 2016) developed an African feminist *methodology* built on women's scholarship and activism. While in other work I have discussed definitions of African feminism, and tenets guiding its methodology, here I examine structural factors operating as mechanisms that impede the development of African feminist thought and decelerate its uptake, namely that encumbered access to information and communications technology (ICT) among women on the African continent and in its Diaspora creates barriers to communal discourse and problem solving. Leaving African women out of the conversation may be partly responsible for the pervasiveness of neoliberalism today. Previously, I have argued that pursuing African feminism provides the opportunity to create intellectual tools to build toward social justice instead of reproducing master narratives (Zerai and Banks 2002). Without the intellectual diversity that African feminism offers, capital expansion and subsequent development, even in the wake of social movements and protest, morphs not into something liberatory, but into another way to reinforce differential access to resources that ideally would be provided as public goods, such as clean water, safe sanitation, reliable and clean energy, and accessible communication technology. For example, in terms of building an infrastructure for development, "public management" is an approach to government and nonprofit administration that resembles private-sector management. In particular, it emphasizes tools that maximize efficiency and effectiveness. However, public management approaches are controversial because "efficiency" is prioritized over meeting the needs of the most marginalized members of society:

> From the end of the 1970s to the 1990s ... governments around the world were engaged in widespread and sustained reforms of their public administration. These reforms were born out of economic recession, but also had political and social drivers. They were initiated by the political apex and fuelled by New Right ideology. Collectively, these reforms came to be termed New Public Management (NPM) ... NPM is characterised by marketisation, privatisation, managerialism, performance measurement and accountability. This employment of corporate attitudes in public administration is grounded on certain theories, mainly public choice, transaction cost analysis and principal–agent theory.
>
> (Tolofari 2005)

In some well-endowed contexts, information technology promises to contribute to enhanced availability of some kinds of knowledge. But in these spaces, I continue to see disparate access, especially along the lines of educational background, language, and age. Some drivers of innovation aim to provide open source materials, while others seek to capitalize on what they see as their intellectual property. In emerging markets, where centralized landline systems are virtually nonexistent, mobile technology has gotten ahead and the principles of neoliberal market-driven expansion trump any efforts to rein in, centrally manage, tax, or ensure nondiscrimination in terms of access to what potentially could be a public good.

African feminist principles present an alternative perspective in line with a people-centered[4] approach to public management that is informed by the values of social justice, equity, diversity, and inclusion, and that eschews heteronormativity, ableism, religious chauvinism, exclusive nationalisms, and marginalization of underrepresented groups from access to technology. Access to technology and especially mobile technology is potentially a major public management issue of the twenty-first century. Without it, members of emerging economies cannot narrow the digital divide.

How can attention to the experiences of African women and postcolonial feminist scholarship provide a unique perspective on ICT, and the wider mobile ecosystem in particular? In my work, I emphasize the voices of women and show how their framings are important when I examine science and technology in society. Women's lack of access to technology and mobile design, and planning for its distribution without considering women or consulting with them leads to fundamental design flaws and slow uptake among women. When technology is inaccessible, digital divides result in both dire consequences for women in African settings and thwart the flow of knowledge within various scholarly communities transnationally.

In order to examine this issue, we need to pay attention to the "wider mobile ecosystem," which includes: industry (mobile network providers, handset manufacturers, and content and application developers), policy-makers, donors, academics, the international development community (GSMA 2015 Report), as well as end users. And the working conditions of those who produce this technology must also be taken into account.

The mobile ecosystem is important to African women because the infrastructure to build and maintain landlines in most of the continent is absent. As is evident in the sample of countries shown in Table 0.2, working landlines are not available in the vast majority of African households. Audre Lorde said that "even the form our creativity takes is often a class issue." She argues for poetry as the most economical art form, compared to prose, requiring "the least physical labor, the least material, and the one which can be done between shifts ... on the subway, and on scraps of surplus paper." In this same way, cellular technology provides a more economical way for families in Africa to communicate with one another across distance.

Table 0.2 Households with working landlines in select African countries

Senegal	6%
South Africa	5%
Kenya	3%
Tanzania	2%
Ghana	1%
Nigeria	1%
Uganda	1%
Median	**2%**

Source: Pew Research Center *Spring 2014 Global Attitudes Survey*, Q7a–b.

Though numbers are growing and are well above those for landlines, there continues to be a gap in access to cell phones in Africa compared to Europe and North America, largely due to their cost. In 2017, Europe and North America still have the highest rates of cell phone users as a percentage of the population, at 85% and 84% unique mobile-cellular subscribers, respectively. Currently, in Sub-Saharan Africa, 44% of the population, representing over 420 million unique subscribers, have mobile-cellular telephone subscriptions,[5] according to the GSMA *The Mobile Economy: Sub-Saharan Africa 2017* and *The Mobile Economy Report 2018*.

Demographic and Health Survey (DHS) data from select countries provide evidence of the range in cell phone ownership by country, from under 10% to upwards of 90% of households with childbearing-aged women own at least one cell phone on the continent. DHS data also show tremendous growth markets in Africa. Cell phone ownership has more than doubled in most countries measured, as shown in Table 0.3.

Table 0.3 Cell phones in women's households and growth in cell phone market since last DHS data collection

Country and year of data collection	*Percentage of women's households owning cell phones*	*Rate increase from time of last DHS*
Egypt 2005	22.6	NA
Egypt 2008	40.7	**18.1**
Egypt 2014	92.1	**51.4**
Ethiopia 2005	7.3	NA
Kenya 2008–2009	64.4	NA
Malawi 2004	5.5	NA
Malawi 2010	43.2	**37.7**
Mozambique 2011	52.9	NA
Nigeria 2008	48.0	NA
Nigeria 2013	78.5	**30.5**
Tanzania 2010	56.6	NA
Uganda 2006	20.3	NA
Uganda 2011	65.6	**45.2**
Zambia 2007	38.3	NA
Zambia 2013	73.6	**35.3**
Zimbabwe 2005–2006	17.1	NA
Zimbabwe 2010–2011	66.6	**49.5**

Source: Zerai's analysis of data extracted by DHS implementing partners and ICF International. Demographic and Health Surveys 2004–2014 (see table for sources). Data extract from DHS Recode files. IPUMS-Demographic and Health Surveys (IPUMS-DHS), version 3.0, Minnesota Population Center and ICF International (Distributors). Accessed at: http://idhsdata.org

However, most cell phones in Africa are *not* smartphones. Pew Research Center data for select African countries shows that about 15% own smartphones, whereas 64% own smartphones in the U.S. (2014). Most nationally representative sample data in Africa provide only household-level cell phone ownership. This provides no sense of *women's* access to phones.

There exists a gender gap in cell phone usage in middle and low-income countries. Because borrowing a phone does not ensure access to a phone when it is needed, and there are gender dimensions in terms of access to cell phones, the current convention is to define "unconnected" women as those who do not individually own phones, but who may possibly borrow one. The majority (59%) of women do not own cell phones in middle and low-income countries. The bulk of women in the world who are unconnected reside in Sub-Saharan Africa, where 64% are unconnected (307 million) and South Asia, where 72% of women are unconnected (594 million), according to the World Bank, and Altai Consulting Analysis (as shown in Figure 4.2 of the GSMA 2015 Report, p. 13). Also, women at the base of the economic pyramid are even less likely to own a cell phone. In Papua New Guinea, for example, 71% of women do not have access to cell phones, and among the poorest, only 16% own one. In one study, 96% say they cannot afford one (GSMA 2014).

According to focus group data, some other barriers to women's access and usage of cell phones in low-income countries include expensive battery charging, fear of having their cell phones stolen, and concerns about cyber harassment (GSMA Connected Women Global Development Alliance 2015, p. 40). And some revealed that such harassment can even create tensions in the household for women who are accused of being willing victims (GSMA 2015, p. 28).

In fact, for some women, cell phone ownership provides a tool that allows the men in their lives to track and control them. In Niger, for example, most (86% of women and 79% of men) agree that it is acceptable for a husband to check the numbers on his wife's cell phone, corresponding to the perception that cell phones can be used to pursue illicit affairs. Notably, during focus group discussions in Niger, some women reported that because of jealousy and fear of adultery, men sometimes change their wife's SIM card after they get married or do not to allow their wives to have a phone at all. Male respondents in an urban focus group echoed these sentiments: "The main reason for women to get a phone is to get men." "For men, the main reason is work" (urban males, Niger; GSMA 2015, p. 32). One of the key reasons for this exertion of "control," cited across many countries, was the perception (among both women and men) that cell phones can be used to pursue romantic liaisons outside of marriage (GSMA 2015, p. 60).

Finally, many women do not have control over the decision to buy cell phones. In a systematic study in a rural Kenyan community, women who have cell phones usually receive them from husbands as second-hand gifts, though they are less likely to receive them than their (usually male) young adult children. And 40% of women who own cell phones have phones *that do not work* (Murphy and Priebe 2011).

The digital divide

The existence of a digital divide has been well documented and is underscored by the United Nations Educational, Scientific and Cultural Organization (UNESCO). The UNESCO and International Social Science Council World Report in 2010 articulated twenty-first-century priorities and expounded the need to address knowledge divides between scholars in the Southern hemisphere and North Atlantic researchers. They explain that Africa is affected by the structural and epistemic disparities that produce the knowledge divides (Harrison 2016). The digital divide separates and marginalizes the Global South, especially women in those settings, and intersectional perspectives help us to understand that digital divides can take place and multiplicatively affect people living in: rural areas; underdeveloped provinces within countries on the African continent (Zerai 2014; Henderson, Zerai, and Morrow 2017); between core and periphery communities within a town such as Mombasa suburbs versus Mombasa informal or refugee settlements; and even, within a family and its household, between men and their (sometimes) victimized wives, given that high percentages of married women experience intimate partner harassment (e.g. control through tracking) and violence (Garćia-Moreno 2005, Devries et al 2013; Cools and Kotsadam 2017).

Consequences of the digital divide

The literature shows that there are multiple consequences of the digital divide for women's lives. These are summarized below.

Consequence 1: Missed opportunities for prenatal care surveillance

The first consequence is missed opportunities for prenatal care surveillance (especially for high-risk pregnancies) and to improve health center use for prenatal care. For example, pregnant Nigerians given free closed users' group cell phones to communicate with health workers were more likely to utilize health centers for prenatal care, deliveries, and primary health care (Oyeyemi and Wynn 2014).

Consequence 2: Missed opportunities for primary health care and social support

Cell phone technology used to send reminder text messages and to monitor microbicide use resulted in correct dosing (Gengiah et al. 2014). Cell phone text reminders have improved adherence to HIV treatment (Gengiah et al. 2014). Streaming health promotion videos on handheld devices resulted in reduced expectations to engage in unprotected sex (Jones 2008).

Lack of communications has an effect on the individual level, but lack of access to ICT also has deleterious effects on communities as well. The first half of Table 0.4 demonstrates that hospitals, normally found only in urban centers, are more likely to have ICT than health centers, dispensaries, clinics, and health posts (that are distributed more widely within countries). And as shown in Table 0.4, there are systematically lower percentages of communications equipment and internet access in public health facilities, to which a broader base of residents have access, compared to private-for profit and even faith-based facilities in some countries.

Consequence 3: Lack of access and difficulties in receiving remittances

Lack of access results in greater difficulties in receiving remittances home from husbands and adult children and inability to connect to families of origin and migrant husbands. A cell phone is an important tool to send remittances home to wives (GSMA 2015; Pew Research Center 2015). In a Mudzi District (Zimbabwe) survey, rural residents use mobile money more to receive money than any other service (Chinakidzwa, Mbengo, and Nyatsambo 2015). And cell phones facilitate connecting families of origin and to migrant husbands (Murphy and Priebe 2011).

Consequence 4: Lack of access to call helplines and technology for women who experience gender-based harassment and violence

For example, in Morocco and Egypt, cyber activists are using digital technologies to combat sexual harassment and violence (Skalli 2014). And free call helplines provide much-needed services to women who experience gender-based violence in South Africa (Raaber 2007).

Table 0.4 Communications and internet access at health facilities (percentage), as reported in recent Service Provision Assessment of the Demographic and Health Surveys (DHS)

Country	*Equipment*	*Facility type*					*Managing authority*		
		Hospital	*Health center*	*Dispensary*	*Clinic*	*Health post*	*Public/ government*	*Private for-profit*	*Faith-based**
Malawi 2015	Communication equipment	90	70	45	85	56	69	87	71
	Computer with internet	73	21	13	49	NA	21	42	44
Senegal 2015	Communication equipment	96	80	NA	NA	39	37	77	NA
	Computer with internet	95	81	NA	NA	35	35	70	NA
Tanzania 2014–2015	Communication equipment	94	62	47	62	NA	47	65	56
	Computer with internet	68	26	8	22	NA	7	31	20

Sources:

Ministry of Health (MoH) (Malawi) and ICF International 2014. Malawi Service Provision Assessment (MSPA) 2013–2014. Lilongwe, Malawi, and Rockville, MD, USA: MoH and ICF International.

Agence Nationale de la Statistique et de la Démographie (ANSD) (Senegal) and ICF International. 2015. Senegal: Enquête Continue sur la Prestation des Services de Soins de Santé (ECPSS) 2014. Rockville, MD, USA: ANSD and ICF International.

Ministry of Health and Social Welfare (MoHSW) (Tanzania Mainland), Ministry of Health (MoH) (Zanzibar), National Bureau of Statistics (NBS), Office of the Chief Government Statistician (OCGS), and ICF International 2015. Tanzania Service Provision Assessment Survey (TSPA) 2014–2015. Dar es Salaam, Tanzania, and Rockville, MD, USA: MoHSW, MoH, NBS, OCGS, and ICF International.

* NA = data is not available.

Note: In Malawi, faith-based facilities are the Christian Hospitals Association of Malawi.

These digital divides underscored by low cell phone ownership impair access to crucial knowledge and networks for women on the margins; *thus, they exacerbate existing gender inequalities.* And *also* importantly, for us whose vocation is the production of knowledge, digital divides cut us off from the perspectives, epistemologies, and wisdoms of those with limited access to ICT.

ICT and governance

In recent work, on the basis of their analysis of World Bank development and governance indicators from 2000 to 2012, Asongu and Nwachukwu (2016) argue that cell phones contribute to a diffusion of knowledge that results in better governance in Sub-Saharan Africa. Their analysis produces robust results that high rates of cell phone ownership lead to subsequent improvements in governance constructs created by Kaufmann et al. (2010), including "rule of law, corruption-control, regulation quality, government effectiveness, political stability/no violence and voice and accountability" (Asongu and Nwachukwu 2016: 135). While this work is compelling, I argue that it does not address gender asymmetry. I endeavor to examine cell phone usage and its impact on education and knowledge diffusion among *women*, and further to analyze the impact of ICT on community-level and national governance factors and values that focus on issues of gender equity.

Breaking down the digital divide

We must make visible that we live our own lives of convenience at the expense of those who are working in poor conditions to produce our devices. We must seek to disrupt the conventional division of labor, in which African communities provide only the raw resources to make technology (West 2016) and markets to buy it (the focus of the GSMA's work). Answering UNESCO's call to work in dialogue with intellectuals and practitioners in the Southern hemisphere is a crucial intervention if we want to stem the process of critical decay within humanities research, the social sciences, and technological innovation that is resulting from their lack of intellectual diversity. For many of us, global themes and essential questions inspire our academic pursuits. Moving beyond our limited understandings requires including marginalized groups. Addressing humanity's essential questions requires a free flow of ideas that eschews the marginalization of women in the Global South. This cannot happen if we continue to reify transnational and gendered digital divides. A reimagining of cell phone design, production, and distribution that places women at the center *and* expanding our intellectual canons to include scholarly works of Afican women may help us to build a stable bridge across digital divides.

A précis of the chapters

In the ensuing chapters, I challenge the current literature that praises the contribution of cell phones to the diffusion of knowledge and resulting better governance in Sub-Saharan Africa. I trouble the androcentrism of this literature and its basis in neoliberal concepts of governance. I then signal my plan to reexamine the data in this literature and to enrich it by bringing in data sources that focus on women's experiences. And finally, I forecast the methodology to be utilized throughout the book that ensures that women scholars' voices are sought and engaged in order to build robust women-centered measures of diffusion of knowledge and good governance.

In Chapter 1, "The mobile ecosystem and internet access on the African continent: asymmetry and the gender digital divide," I analyze DHS and World Bank data to describe differences between regions of the African continent and between nations in relation to internet access and cell phone ownership. I also examine available evidence concerning the gender digital divide.

In Chapter 2, "ICT, women's status and governance in Zimbabwe," I make the case for a woman-centered methodology, build a database of the works of female and gender-conscious scholars and activists in Zimbabwe, and from it delineate locally determined values that focus on issues of gender equity and good governance. Next, I build a conceptual model to analyze the impact of ICT on these factors. Chapter 3, "ICT, women's status and governance in Tanzania, 2010 and 2015–2016," is similar to the preceding, except that I examine organic definitions of gender equity and good governance from a Tanzania universities-centered female and gender-conscious scholar and activist database. I also analyze data from the Tanzania DHS in 2010 and 2015–2016. Chapter 4, "ICT, women's status and governance in Malawi, 2010 and 2015–2016," follows in this vein with a focus on women scholars and activists in Malawi and by analyzing Malawi's DHS data from 2010 and 2015–2016.

In Chapter 5, "ICT, diffusion of knowledge to women, gender-inclusive governance, and impacts on women's lives in three African nations," I merge DHS data with World Bank development and governance indicators in order to test for lag effects on the availability of mobile technology in women's households and on the diffusion of knowledge, governance, and gender equity. I challenge the argument that cell phones contribute to the diffusion of knowledge, resulting in better governance in Sub-Saharan Africa, noting that we cannot fully understand "better governance" without also considering factors that are relevant to gender equity.

After examining ICT and people-centered governance (to include measures of gender status) in Chapter 5, in the conclusion I discuss the relevance of the findings to various intellectual communities and communities of practice. These include studies of science and technology in society, development effectiveness, public administration of ICT, and the study of women and gender and governance in African countries.

Notes

1 Portions of this introduction were originally presented as my keynote address, titled "In the Traditions of Professor Victor C. Uchendu and Professor Ifi Amadiume, A Proposal for Expanding African Transnational Feminist Praxis: Creating a Database of Women's Scholarship Concerning Women's Health in Zimbabwe", at the symposium to commemorate the 50th anniversary of the release of Professor Victor Uchendu's *The Igbo of Southeastern Nigeria.* May 2, 2016, at Virginia Polytechnic Institute and State University.

2 See Cooper 1998 edited by Lemert and Bhan for an anthology of her writings from the 1890s.

3 Hereinafter, I will refer to this field as "African feminism," in honor of a pan-African perspective that includes both Africa and its Diaspora.

4 The term "people-centered" governance has been taken up by ActionAid International in their quest to "achieve a safer world for poor and excluded people" (2005). The idea of citizen-focused governance, especially in global cities is also engaged in the work of Horace Campbell (2003), Saskia Sassen (2003, 2005, 2017), Asef Bayat (2012) and the UN Compendium of Innovative Practices in Public Governance and Administration for Sustainable Development (2016). While I do not expound on the definition of people-centered governance, I believe that its principles are consistent with humanistic social justice oriented African feminist perspectives.

5 According to the UN, mobile-cellular subscriptions are "defined as the number of mobile-cellular telephone subscriptions per 100" in the population. "Mobile-cellular telephone subscriptions refers to the number of subscriptions to a public mobile-telephone service that provide access to the PSTN using cellular technology" (accessed at: https://millenniumindicators.un.org/unsd/mdg/Metadata.aspx?IndicatorId=0&SeriesId=780).

1 The mobile ecosystem and internet access on the African continent

Asymmetry and the gender digital divide

In this chapter, in preparation of my reexamination of the ICT–knowledge diffusion hypothesis offered by Asongu and Nwachukwu (2016), I examine World Bank indicators and Demographic and Health Surveys data to describe differences between nations on the African continent in relation to cell phone ownership and internet access. I also examine the available evidence concerning the gender digital divide. An overview of countries for which cell phone ownership data is available provides a useful backdrop to examining the experiences of women in Malawi, Tanzania, and Zimbabwe in later chapters.

World Bank indicators: differences among nations of the African continent in relation to internet access and cell phone ownership and the gender digital divide

World Bank indicators provide data on cell phone ownership in all countries on the African continent for which this data has been collected. Out of 54 countries, data on the mobile-cellular telephone subscriptions per 100 people is available for all nations. And data on mobile-cellular telephone accounts owned by women is available in most countries (34 out of 54). Mobile-cellular telephone subscriptions per 100 people ranged from 3 to 180 in 2005 and ranged from 7 to 169 in 2015 (latest data available). And a minority of women (aged 15 and older) own cell phones subscriptions. This percentage ranges from .66 (less than 1%) to 55%. See Table 1.1.

In relation to internet users (per 100 people), in 2010 numbers ranged as low as low as .6 per 100 to a high of up to 58 per 100 in 2015. Increases in internet users per 100 people from 2010 to 2015 are found across all 53 countries for which comparative data are available. Similar patterns are found in terms of the number of secure servers, as well as number of secure servers per 1,000,000 people, with an increase in secure servers

Table 1.1 Mobile-cellular telephone subscriptions (2010 and 2015) and mobile-cellular telephone accounts owned by women (2014)

Country	*Mobile-cellular subscriptions* (per 100 people), 2010*	*Mobile-cellular subscriptions (per 100 people), 2015*	*Mobile account, female (% age 15+), 2014*
Algeria	88.44	113.03	NA
Angola	48.10	60.84	NA
Benin	74.40	85.64	1.15
Botswana	120.01	169.00	19.46
Burkina Faso	36.73	80.64	3.06
Burundi	18.17	46.22	0.66
Cabo Verde	76.27	127.15	NA
Cameroon	41.88	71.85	2.09
Central African Republic	22.51	20.45	NA
Chad	24.53	40.17	4.43
Comoros	24.20	54.80	NA
Congo, Democratic Republic of the	19.01	52.99	7.43
Congo, Republic of the	90.44	111.66	2.53
Côte d'Ivoire	82.20	119.31	20.03
Djibouti	19.86	34.68	NA
Egypt	90.50	110.99	0.13
Equatorial Guinea	57.36	66.72	NA
Eritrea	3.23	7.05	NA
Ethiopia	7.87	42.76	0.05
Gabon	103.46	168.92	6.14
Gambia	87.96	131.26	NA
Ghana	71.87	129.74	11.84
Guinea	36.78	87.17	0.58
Guinea-Bissau	42.69	69.27	NA
Kenya	61.03	80.68	54.90
Lesotho	49.15	105.52	NA
Liberia	39.70	81.09	NA
Libya	180.45	157.00	NA
Madagascar	36.58	46.02	4.23
Malawi	20.76	35.34	2.53
Mali	53.20	139.61	8.99
Mauritania	76.91	89.32	6.62
Mauritius	96.77	140.57	0.17
Morocco	101.07	126.87	NA
Mozambique	30.14	74.24	NA
Namibia	89.50	102.10	8.40
Niger	23.08	46.50	2.17
Nigeria	54.66	82.19	2.07
Rwanda	32.75	70.48	16.13

Country	*Mobile-cellular subscriptions* (per 100 people), 2010*	*Mobile-cellular subscriptions (per 100 people), 2015*	*Mobile account, female (% age 15+), 2014*
São Tomé and Principe	57.64	65.09	NA
Senegal	64.43	99.95	4.72
Seychelles	128.92	158.12	NA
Sierra Leone	34.77	89.53	3.59
Somalia	6.73	52.47	32.06
South Africa	97.90	159.27	13.89
South Sudan	14.45	23.86	NA
Sudan	41.54	70.53	NA
Swaziland	60.83	73.20	NA
Tanzania	46.66	75.86	26.58
Togo	41.27	64.95	1.32
Tunisia	104.54	129.93	0.38
Uganda	37.74	50.37	29.04
Zambia	41.21	74.47	9.48
Zimbabwe	58.88	84.79	19.10

Sources: Demirgüç-Kunt et al., 2015, Global Financial Inclusion Database, World Bank; International Telecommunication Union, World Telecommunication/ICT Development Report and database.

* "Mobile-cellular account" denotes the percentage of respondents who report personally using a cell phone to pay bills or to send or receive money through a GSM Association (GSMA) Mobile Money for the Unbanked (MMU) service in the past 12 months; or receiving wages, government transfers, or payments for agricultural products through a cell phone in the past 12 months. Mobile-cellular telephone subscriptions are subscriptions to a public mobile telephone service that provide access to the PSTN using cellular technology. The indicator includes (and is split into) the number of postpaid subscriptions and the number of active prepaid accounts (i.e. that have been used during the last three months). The indicator applies to all mobile cellular subscriptions that offer voice communications. It excludes subscriptions via data cards or USB modems, subscriptions to public mobile data services, private trunked mobile radio, telepoint, radio paging, and telemetry services.

Note: NA: women's mobile-cellular telephone account percentages are not available.

and servers per 1,000,000 people from 2010 to 2015 in all countries for which comparative data are available. See Table 1.2.

DHS data and factors associated with a cell phone in households

In this section, I examine data from the Demographic and Health Surveys (DHS). With support from the United States Agency on International Development (USAID), DHS staff work in cooperation

Table 1.2 Internet users per 100 people, secure internet servers, and servers per 1 million people (2010 and 2015)

Country	*Internet users (per 100 people), 2010*	*Internet users (per 100 people), 2015*	*Secure internet servers, 2010*	*Secure internet servers, 2015*	*Secure internet servers (per 1 million people), 2010*	*Secure internet servers (per 1 million people), 2015*
Algeria	12.5	38.20	31	101	0.86	2.55
Angola	2.8	12.40	52	137	2.45	5.48
Benin	3.1	6.79	1	42	0.11	3.86
Botswana	6.0	27.50	17	39	8.30	17.24
Burkina Faso	2.4	11.39	4	12	0.26	0.66
Burundi	1.0	4.87	2	6	0.21	0.54
Cabo Verde	30.0	43.02	7	27	14.27	51.87
Cameroon	4.3	20.68	11	58	0.53	2.48
Central African Republic	2.0	4.56	NA	NA	NA	NA
Chad	1.7	2.70	NA	3	NA	0.21
Comoros	5.1	7.46	1	NA	1.43	NA
Congo, Dem. Rep. of the	0.7	3.80	8	27	0.12	0.35
Congo, Rep. of the	5.0	7.62	5	8	1.23	1.73
Côte d'Ivoire	2.7	21.00	21	93	1.04	4.10
Djibouti	6.5	11.92	5	8	6.02	9.01
Egypt	21.6	35.90	186	498	2.27	5.44
Equatorial Guinea	6.0	21.32	2	3	2.74	3.55
Eritrea	0.6	1.08	NA	NA	NA	NA
Ethiopia	0.8	11.60	11	24	0.13	0.24
Gabon	7.2	23.50	12	17	7.78	9.85
Gambia	9.2	17.12	5	11	2.95	5.53
Ghana	7.8	23.48	42	137	1.73	5.00
Guinea	1.0	4.70	4	6	0.36	0.48
Guinea-Bissau	2.5	3.54	1	3	0.61	1.63
Kenya	14.0	45.62	105	421	2.60	9.14
Lesotho	3.9	16.07	1	9	0.50	4.22
Liberia	2.3	5.90	3	16	0.76	3.55
Libya	14.0	19.02	9	24	1.44	3.82
Madagascar	1.7	4.17	10	40	0.47	1.65
Malawi	2.3	9.30	4	23	0.27	1.34
Mali	2.0	10.34	11	22	0.73	1.25
Mauritania	4.0	15.20	7	11	1.95	2.70
Mauritius	28.3	50.14	111	222	88.77	175.83
Morocco	52.0	57.08	95	212	2.96	6.17

Country	*Internet users (per 100 people), 2010*	*Internet users (per 100 people), 2015*	*Secure internet servers, 2010*	*Secure internet servers, 2015*	*Secure internet servers (per 1 million people), 2010*	*Secure internet servers (per 1 million people), 2015*
Mozambique	4.2	9.00	20	61	0.82	2.18
Namibia	11.6	22.31	32	70	14.59	28.47
Niger	0.8	2.22	3	5	0.18	0.25
Nigeria	24.0	47.44	191	474	1.20	2.60
Rwanda	8.0	18.00	7	48	0.68	4.13
São Tomé and Principe	18.8	25.82	3	2	17.56	10.51
Senegal	8.0	21.69	14	79	1.08	5.22
Seychelles	41.0	58.12	100	45	1,113.96	484.39
Sierra Leone	0.6	2.50	1	6	0.17	0.93
Somalia	NA	1.76	1	1	0.10	0.09
South Africa	24.0	51.92	3,129	7,143	61.63	129.97
South Sudan	7.0	17.93	NA	3	NA	0.24
Sudan	16.7	26.61	1	3	0.02	0.07
Swaziland	11.0	30.38	12	20	10.06	15.54
Tanzania	2.9	5.36	18	109	0.39	2.04
Togo	3.0	7.12	12	46	1.88	6.30
Uganda	12.5	19.22	29	75	0.87	1.92
Zambia	10.0	21.00	16	70	1.15	4.32
Zimbabwe	6.4	16.36	13	103	0.93	6.60

Notes:
Data from database: World Development Indicators.

Last updated: February 1, 2017.

Accessed at: http://databank.worldbank.org/data/reports.aspx?source=2&series=IT.CEL.SETS.P2&country=#

NA = data is not available.

with health and statistics agencies within a majority of African countries to gather sample data that are representative of each country's population. DHS data focus on factors relevant to maternal and child health, though cell phone ownership is one of the newer variables that has been added to the DHS. While DHS data are imperfect, they provide an overview, on the basis of nationally representative samples, of the availability of cell phones in the households of women aged 15–49. Unlike the World Bank Indicators, DHS data are available at the individual and household levels, so I can examine factors associated with a cell phone in households.

DHS descriptive data

Among 19 countries for which information on household cell phone ownership is collected by the DHS (at one point in time or over multiple points in time), an average of 51.3% of households owned a cell phone (either currently or as an aggregate over up to three time periods), ranging from 6% to 86% of households surveyed. See Table 1.3.

The marital status and/or relationship to the household head of the respondent are important because they provide some sense of who may be making decisions about the purchase of cell phones. In the representative sample of childbearing-aged women (aged 15 to 49), from 31% to 83% of respondents are wives in households headed by men (the mean is 57%). A minority, from 3% to 29% of respondents, are female heads of households (the mean is 11%). See Table 1.4.

Table 1.5 shows liberalization on matters concerning gender over time in many countries, namely that fewer agree that wife beating is justified for refusing sex in latter survey years. In contrast, countries where the difference over time is not likely to be statistically significant include Tanzania (slight decrease from 28% of female respondents agreeing that wife beating is justified if woman refuses to have sex in 2004 to 27% agreeing in 2010 DHS), Uganda (increase in agreement that wife beating is justified to 31% in 2006, though there was a 10-point decrease in the percentage of women agreeing in 2011), Zimbabwe (increase to 24% in 2005–2006, though there was a 8-point decrease in 2010–2011), Rwanda (increase in 2010 to 36%), Mali (increase in 2012 to 58%), Madagascar (increase in 2008 to 10%), Guinea (increase in 2012 to 70%), and Benin (increase in 2006 to 18%, though there was an 11-point decrease in 2011). Even more extreme, note that in Guinea, Mali and Niger, the majority of household respondents (largely wives in male-headed households) *agreed* that wife beating is justified for refusing sex.

Table 1.6 shows that distance to a health facility to attend to women's health needs is "a big problem" for respondents in an average (mean) of 34% of households (ranging from 7-63%). For most countries, with multiple years assessed, the perception that distance to a facility is "a big problem" for women obtaining their own medical health care erodes over time (with the exception of Benin, Ghana, Ethiopia, and Egypt).

Finally, in Table 1.7, 39% of women interviewed had no education and 34.1% had only primary-level education. So, the majority (73.5%) had primary-level or less education.

These data describe the "average" experience for most women captured by the DHS in the 19 African countries for which cell phone

Table 1.3 Crosstabulation table of DHS country/year and whether the respondent's household has a cell phone in 19 countries on the African continent

Country and year	*Yes (count)*	*Yes (%)*	*Missing (count)*	*Total (count)*
Benin 2006	4,410	24.8	30	17,794
Benin 2011	12,573	75.7	0	16,599
Burkina Faso 2010	11,354	66.4	3	17,087
Cameroon 2004	3,026	28.4	21	10,656
Cameroon 2011	11,359	73.6	19	15,426
Côte d'Ivoire 2011	8,637	85.9	3	10,060
Egypt 2005	4,409	22.6	17	19,474
Egypt 2008	6,724	40.7	24	16,527
Egypt 2014	20,047	92.1	3	21,762
Ethiopia 2005	1,025	7.3	10	14,070
Guinea 2012	6,535	71.5	3	9,142
Kenya 2008–2009	5,440	64.4	1	8,444
Madagascar 2008	5,743	33.1	6	17,375
Malawi 2004	645	5.5	2	11,698
Malawi 2010	9,955	43.2	18	23,020
Mali 2012	8,381	80.4	0	10,424
Mozambique 2011	7,274	52.9	0	13,745
Niger 2006	1,596	17.3	6	9,223
Niger 2012	7,006	62.8	5	11,160
Nigeria 2008	16,016	48.0	65	33,385
Nigeria 2013	30,567	78.5	73	38,948
Rwanda 2005	983	8.7	38	11,321
Rwanda 2010	6,469	47.3	2	13,671
Tanzania 2010	5,738	56.6	4	10,139
Uganda 2006	1,729	20.3	1	8,531
Uganda 2011	5,688	65.6	0	8,674
Zambia 2007	2,735	38.3	2	7,146
Zambia 2013	12,081	73.6	6	16,411
Zimbabwe 2005–2006	1,523	17.1	11	8,907
Zimbabwe 2010–2011	6,107	66.6	0	9,171
Zimbabwe 2015	6918	69.5	0	9955
Total	225,775	51.3	373	439,990

Citation: Elizabeth Heger Boyle, Miriam King, and Matthew Sobek. IPUMS-Demographic and Health Surveys: Version 4.1 (dataset). Minnesota Population Center and ICF International, 2017. http://doi.org/10.18128/D080.V4.0.

Source: ICF: The DHS Model Datasets 2017 (dataset), Rockville, MD, USA: ICF. Data extract from all datasets noted above. IPUMS-Demographic and Health Surveys (IPUMSDHS), version 4.1, Minnesota Population Center and ICF International (Distributors). Accessed at: http:// idhsdata.org

Table 1.4 Crosstabulation table of country and year of data collection by relationship to household head

	Relationship to household head			
	Head	*Wife*	*Other*	*Total*
Benin 1996	318 5.8%	3,253 59.2%	1,920 35%	5,491
Benin 2001	559 9%	3,551 57.1%	2,109 33.9%	6,219
Benin 2006	1,947 10.9%	10,892 61.2%	4,955 27.8%	17,794
Benin 2011	1,787 10.8%	10,000 60.2%	4,812 29%	16,599
Burkina Faso 1993	219 3.4%	4,104 64.6%	2,031 32%	6,354
Burkina Faso 1998	216 3.4%	4,322 67.1%	1,907 29.6%	6,445
Burkina Faso 2003	437 3.5%	8,021 64.3%	4,019 32.2%	12,477
Burkina Faso 2010	845 4.9%	11,901 69.6%	4,341 25.4%	17,087
Cameroon 1991	307 7.9%	2,041 52.7%	1,523 39.3%	3,871
Cameroon 1998	534 9.7%	2,484 45.2%	2,483 45.1%	5,501
Cameroon 2004	1,182 11.1%	4,988 46.8%	4,486 42.1%	10,656
Cameroon 2011	1,693 11%	6,872 44.5%	6,861 44.5%	15,426
Côte d'Ivoire 1994	488 6%	3,424 42.3%	4,187 51.7%	8,099
Côte d'Ivoire 1998	168 5.5%	1,198 39.4%	1,674 55.1%	3,040
Côte d'Ivoire 2011	760 7.6%	4,814 47.9%	4,486 44.6%	10,060
Egypt 1992	448 4.5%	7,126 72.2%	2,290 23.2%	9,864
Egypt 1995	731 4.9%	10,478 70.9%	3,570 24.2%	14,779
Egypt 2000	749 4.8%	11,531 74%	3,293 21.1%	15,573
Egypt 2005	798 4.10%	14,423 74.1%	4,253 21.8%	19,474
Egypt 2008	815 4.9%	12,492 75.6%	3,220 19.5%	16,527
Egypt 2014	804 3.7%	18,064 83%	2,894 13.3%	21,762

	Relationship to household head			
	Head	Wife	Other	Total
Ethiopia 2000	2,035 13.2%	7,590 49.4%	5,742 37.4%	15,367
Ethiopia 2005	1,872 13.3%	7,249 51.5%	4,949 35.2%	14,070
Ethiopia 2011	2,617 15.8%	8,010 48.5%	5,888 35.7%	16,515
Ghana 1993	1,326 29.1%	2,030 44.5%	1,206 26.4%	4,562
Ghana 1998	1,140 23.5%	2,191 45.2%	1,512 31.2%	4,843
Ghana 2003	1,003 17.6%	2,594 45.6%	2,094 36.8%	5,691
Ghana 2008	1,045 21.3%	2,108 42.9%	1,763 35.9%	4,916
Guinea 1999	303 4.5%	4,225 62.6%	2,225 32.9%	6,753
Guinea 2005	460 5.8%	4,972 62.5%	2,522 31.7%	7,954
Guinea 2012	513 5.6%	5,277 57.7%	3,352 36.7%	9,142
Kenya 1993	1,412 18.7%	3,088 41%	3,040 40.3%	7,540
Kenya 1998	1,491 18.9%	3,365 42.7%	3,025 38.4%	7,881
Kenya 2003	1,569 19.1%	3,628 44.3%	2,998 36.6%	8,195
Kenya 2008–2009	1,843 21.8%	3,618 42.8%	2,983 35.3%	8,444
Madagascar 1992	672 10.7%	3,033 48.5%	2,555 40.8%	6,260
Madagascar 1997	801 11.3%	3,791 53.7%	2,468 35%	7,060
Madagascar 2003	1,015 12.8%	4,425 55.7%	2,509 31.6%	7,949
Madagascar 2008	2,212 12.7%	9,580 55.1%	5,583 32.1%	17,375
Malawi 1992	595 12.3%	2,886 59.5%	1,368 28.2%	4,849
Malawi 2000	2,000 15.1%	7,468 56.5%	3,752 28.4%	13,220
Malawi 2004	1,717 14.7%	7,130 61%	2,851 24.4%	11,698
Malawi 2010	3,687 16%	12,692 55.1%	6,641 28.8%	23,020

(*continued*)

Table 1.4 (cont.)

	Relationship to household head			
	Head	*Wife*	*Other*	*Total*
Mali 1995–1996	292	7,153	2,259	9,704
	3%	73.7%	23.3%	
Mali 2001	790	9,334	2,725	12,849
	6.1%	72.6%	21.2%	
Mali 2006	797	10,109	3,677	14,583
	5.5%	69.3%	25.2%	
Mali 2012	536	7,634	2,254	10,424
	5.1%	73.2%	21.6%	
Mozambique 1997	1,353	4,620	2,806	8,779
	15.4%	52.6%	32%	
Mozambique 2003	1,743	6,138	4,537	12,418
	14%	49.4%	36.5%	
Mozambique 2011	3,201	5,965	4,579	13,745
	23.3%	43.4%	33.3%	
Niger 1992	244	4,319	1,940	6,503
	3.8%	66.4%	29.8%	
Niger 1998	442	4,801	2,334	7,577
	5.8%	63.4%	30.8%	
Niger 2006	838	5,417	2,968	9,223
	9.1%	58.7%	32.2%	
Niger 2012	829	8,032	2,299	11,160
	7.40%	72%	20.6%	
Nigeria 1990	564	5,703	2,514	8,781
	6.4%	64.9%	28.6%	
Nigeria 1999	643	5,007	4,160	9,810
	6.6%	51%	42.4%	
Nigeria 2003	571	4,449	2,600	7,620
	7.5%	58.4%	34.1%	
Nigeria 2008	3,005	20,990	9,390	33,385
	9%	62.9%	28.1%	
Nigeria 2013	3,574	23,511	11,863	38,948
	9.2%	60.4%	30.5%	
Rwanda 1992	566	3,435	2,550	6,551
	8.6%	52.4%	38.9%	
Rwanda 2000	2,065	4,256	4,100	10,421
	19.8%	40.8%	39.3%	
Rwanda 2005	1,893	4,823	4,605	11,321
	16.7%	42.6%	40.7%	
Rwanda 2010	2,110	5,974	5,587	13,671
	15.4%	43.7%	40.9%	
Tanzania 1991	696	4,645	3,897	9,238
	7.5%	50.3%	42.2%	
Tanzania 1996	848	4,185	3,087	8,120
	10.4%	51.5%	38%	

	Relationship to household head			
	Head	*Wife*	*Other*	*Total*
Tanzania 1999	384 9.5%	1,892 47%	1,753 43.5%	4,029
Tanzania 2004	1,164 11.3%	5,232 50.7%	3,933 38.1%	10,329
Tanzania 2010	1,068 10.5%	5,032 49.6%	4,039 39.8%	10,139
Uganda 1995	1,120 15.8%	3,828 54.1%	2,122 30%	7,070
Uganda 2001	1,312 18.1%	3,816 52.7%	2,118 29.2%	7,246
Uganda 2006	1,570 18.4%	4,114 48.2%	2,847 33.4%	8,531
Uganda 2011	1,590 18.3%	4,078 47%	3,006 34.7%	8,674
Zambia 1992	484 6.9%	3,764 53.3%	2,812 39.8%	7,060
Zambia 1996	839 10.5%	4,009 50%	3,173 39.6%	8,021
Zambia 2001	845 11%	4,011 52.4%	2,802 36.6%	7,658
Zambia 2007	954 13.4%	3,694 51.7%	2,498 35%	7,146
Zambia 2013	2,264 13.8%	8,236 50.2%	5,911 36%	16,411
Zimbabwe 1994	1,192 19.5%	2,221 36.2%	2,715 44.3%	6,128
Zimbabwe 1999	1,244 21.1%	2,168 36.7%	2,495 42.2%	5,907
Zimbabwe 2005–2006	1,857 20.8%	3,157 35.4%	3,893 43.7%	8,907
Zimbabwe 2010–2011	2,604 28.4%	2,872 31.3%	3,695 40.3%	9,171
Zimbabwe 2015	2,437 24.5%	3,676 36.9%	3,842 38.6%	9,955
Total	95,224 11%	489,553 56.7%	278,883 32.3%	863,660

Citation: Elizabeth Heger Boyle, Miriam King, and Matthew Sobek. IPUMS-Demographic and Health Surveys: Version 4.1 (dataset). Minnesota Population Center and ICF International, 2017. http://doi.org/10.18128/D080.V4.0.

Source: ICF: The DHS Model Datasets 2017 (dataset), Rockville, MD, USA: ICF. Data extract from all datasets noted above. IPUMS-Demographic and Health Surveys (IPUMSDHS), version 4.1, Minnesota Population Center and ICF International (Distributors). Accessed at: http://idhsdata.org

Table 1.5 Crosstabulation table of country and survey year by female respondent's attitude to "wife beating"

	Attitude to "wife beating": justified if woman refuses to have sex			
	No	*Yes*	*Don't know*	*Total*
Benin 2001	4,961 79.8%	1,079 17.4%	174 2.8%	6,219
Benin 2006	14,050 79%	3,238 18.2%	487 2.7%	17,794
Benin 2011	15,194 91.5%	1,092 6.6%	313 1.9%	16,599
Burkina Faso 2003	7,031 56.4%	4,688 37.6%	756 6.1%	12,477
Burkina Faso 2010	13,838 81%	3,148 18.4%	94 0.6%	17,087
Cameroon 2004	8,144 76.4%	2,029 19%	480 4.5%	10,656
Cameroon 2011	12,973 84.1%	2,130 13.8%	309 2%	15,426
Côte d'Ivoire 2011	7,450 74.1%	2,396 23.8%	209 2.1%	10,060
Egypt 2005	12,170 62.5%	6,958 35.7%	311 1.6%	19,474
Egypt 2008	12,276 74.3%	4,102 24.8%	126 0.8%	16,527
Egypt 2014	17,357 79.8%	4,098 18.8%	306 1.4%	21,762
Ethiopia 2000	7,779 50.6%	6,804 44.3%	776 5%	15,367
Ethiopia 2005	8,155 58%	5,313 37.8%	594 4.2%	14,070
Ethiopia 2011	10,099 61.2%	6,073 36.8%	328 2%	16,515
Ghana 2003	4,391 77.2%	1,227 21.6%	71 1.2%	5,691
Ghana 2008	4,082 83%	693 14.1%	141 2.9%	4,916
Guinea 2005	2,508 31.5%	4,984 62.7%	455 5.7%	7,954
Guinea 2012	2,621 28.7%	6,437 70.4%	82 0.9%	9,142
Kenya 2003	5,301 64.7%	2,433 29.7%	448 5.5%	8,195
Kenya 2008–2009	6,192 73.3%	1,912 22.6%	335 4%	8,444
Madagascar 2003	7,321 92.1%	442 5.6%	184 2.3%	7,949

	Attitude to "wife beating": justified if woman refuses to have sex			
	No	*Yes*	*Don't know*	*Total*
Madagascar 2008	15,121	1,718	515	17,375
	87%	9.9%	3%	
Malawi 2000	10,429	2,456	333	13,220
	78.9%	18.6%	2.5%	
Malawi 2004	9,728	1,644	321	11,698
	83.2%	14.1%	2.7%	
Malawi 2010	21,377	1,425	208	23,020
	92.9%	6.2%	0.9%	
Mali 2001	2,961	9,443	438	12,849
	23%	73.5%	3.4%	
Mali 2006	6,111	7,862	586	14,583
	41.9%	53.9%	4%	
Mali 2012	4,125	6,047	252	10,424
	39.6%	58%	2.4%	
Mozambique 2003	8,330	4,072	0	12,418
	67.1%	32.8%	0%	
Mozambique 2011	12,776	717	252	13,745
	93%	5.2%	1.8%	
Niger 2006	3,732	5,172	305	9,223
	40.5%	56.1%	3.3%	
Niger 2012	5,790	4,870	495	11,160
	51.9%	43.6%	4.4%	
Nigeria 2003	4,635	2,684	286	7,620
	60.8%	35.2%	3.8%	
Nigeria 2008	22,800	8,952	1,528	33,385
	68.3%	26.8%	4.6%	
Nigeria 2013	30,151	7,589	1,157	38,948
	77.4%	19.5%	3%	
Rwanda 2000	6,314	3,272	833	10,421
	60.6%	31.4%	8%	
Rwanda 2005	9,029	1,513	776	11,321
	79.8%	13.4%	6.9%	
Rwanda 2010	8,378	4,890	401	13,671
	61.3%	35.8%	2.9%	
Tanzania 2004	7,079	2,872	373	10,329
	68.5%	27.8%	3.6%	
Tanzania 2010	7,093	2,701	343	10,139
	70%	26.6%	3.4%	
Uganda 2001	5,193	1,639	414	7,246
	71.7%	22.6%	5.7%	
Uganda 2006	5,563	2,601	363	8,531
	65.2%	30.5%	4.3%	

(*continued*)

Table 1.5 (cont.)

	Attitude to "wife beating": justified if woman refuses to have sex			
	No	*Yes*	*Don't know*	*Total*
Uganda 2011	6,631 76.4%	1,794 20.7%	243 2.8%	8,674
Zambia 2001	3,599 47%	3,692 48.2%	364 4.8%	7,658
Zambia 2007	4,177 58.5%	2,556 35.8%	409 5.7%	7,146
Zambia 2013	10,533 64.2%	5,167 31.5%	702 4.3%	16,411
Zimbabwe 1999	4,378 74.1%	1,318 22.3%	210 3.6%	5,907
Zimbabwe 2005–2006	6,542 73.4%	2,163 24.3%	197 2.2%	8,907
Zimbabwe 2010–2011	7,453 81.3%	1,473 16.1%	245 2.7%	9,171
Zimbabwe 2015	1,444 14.5%	8,334 83.7%	177 1.8%	9,955
Total	433,921 69.1%	173,578 27.7%	19,528 3.1%	627,524

Citation: Elizabeth Heger Boyle, Miriam King, and Matthew Sobek. IPUMS-Demographic and Health Surveys: Version 4.1 (dataset). Minnesota Population Center and ICF International, 2017. http://doi.org/10.18128/D080.V4.0.

Source: ICF: The DHS Model Datasets 2017 (dataset), Rockville, MD, USA: ICF. Data extract from all datasets noted above. IPUMS-Demographic and Health Surveys (IPUMSDHS), version 4.1, Minnesota Population Center and ICF International (Distributors). Accessed at: http:// idhsdata.org

ownership can be analyzed as a variable. The average woman lives in a household where a cell phone is owned (the mean is 51%) by a (usually male) household member. And women's households today are more likely to have cell phones compared to five or ten years ago. Women in households among the 19 countries are more likely to be married and living in a household in which their husbands are considered the head of the household (the mean is 57%). Women in a majority of households do not agree that wife beating is justified, though a large minority (28%) feel it is justified if a wife refuses to have sex. Over one-third (34%) of women have poor access to health facilities to attend to their health needs due to distance. And the majority of women aged 15–49 have primary-level education or lower.

Table 1.6 Crosstabulation table of country/year by whether distance is a barrier to female respondent's health care access

	*Barrier to woman's health care: distance to facility**			
	Not a big problem	*No problem at all*	*Small problem*	*Big problem*
Benin 2001	0	3,740	1,101	1,358
	0%	60.1%	17.7%	21.8%
Benin 2006	10,883	0	0	6,882
	61.2%	0%	0%	38.7%
Benin 2011	9,174	0	0	7,425
	55.3%	0%	0%	44.7%
Burkina Faso 2003	6,807	0	0	5,666
	54.6%	0%	0%	45.4%
Burkina Faso 2010	9,649	0	0	7,427
	56.5%	0%	0%	43.5%
Cameroon 2004	6,523	0	0	4,117
	61.2%	0%	0%	38.6%
Côte d'Ivoire 2011	5,934	0	0	4,102
	59.0%	0%	0%	40.8%
Egypt 2000	13,338	0	0	2,235
	85.6%	0%	0%	14.4%
Egypt 2005	16,141	0	0	3,326
	82.9%	0%	0%	17.1%
Egypt 2008	13,293	0	0	3,205
	80.4%	0%	0%	19.4%
Egypt 2014	17,551	0	0	4,211
	80.6%	0%	0%	19.4%
Ethiopia 2000	14,227	0	0	1,140
	92.6%	0%	0%	7.4%
Ethiopia 2005	5,609	0	0	8,456
	39.9%	0%	0%	60.1%
Ethiopia 2011	6,090	0	0	10,412
	36.9%	0%	0%	63.0%
Ghana 2003	3,600	0	0	2,088
	63.3%	0%	0%	36.7%
Ghana 2008	3,462	0	0	1,445
	70.4%	0%	0%	29.4%
Guinea 2005	3,625	0	0	4,317
	45.6%	0%	0%	54.3%
Madagascar 2003	0	3,891	1,378	2,672
	0%	48.9%	17.3%	33.6%
Madagascar 2008	4,925	0	0	3,615
	28.3%	0%	0%	20.8%

(*continued*)

Table 1.6 (cont.)

	*Barrier to woman's health care: distance to facility**			
	Not a big problem	*No problem at all*	*Small problem*	*Big problem*
Mali 2001	7,536 58.7%	0 0%	0 0%	5,291 41.2%
Mali 2006	8,505 58.3%	0 0%	0 0%	6,051 41.5%
Mali 2012	6,769 64.9%	0 0%	0 0%	3,655 35.1%
Niger 2012	6,424 57.6%	0 0%	0 0%	4,724 42.3%
Rwanda 2000	0 0%	4,744 45.5%	1,599 15.3%	4,071 39.1%
Rwanda 2005	6,844 60.5%	0 0%	0 0%	4,460 39.4%
Rwanda 2010	10,165 74.4%	0 0%	0 0%	3,502 25.6%
Zambia 2013	10,266 62.6%	0 0%	0 0%	6,126 37.3%
Zimbabwe 2015	3,313 33.3%	0 0%	0 0%	6,642 66.7%
Total	207,340 58.4%	12,375 3.5%	4,078 1.1%	121,979 34.4%

Citation: Elizabeth Heger Boyle, Miriam King, and Matthew Sobek. IPUMS-Demographic and Health Surveys: Version 4.1 (dataset). Minnesota Population Center and ICF International, 2017. http://doi.org/10.18128/D080.V4.0.

Source: ICF: The DHS Model Datasets 2017 (dataset), Rockville, MD, USA: ICF. Data extract from all datasets noted above. IPUMS-Demographic and Health Surveys (IPUMSDHS), version 4.1, Minnesota Population Center and ICF International (Distributors). Accessed at: http://idhsdata.org

* According to the IDHS, 'The universe is the population at risk of having a response for the variable in question. In most cases these are the households or persons to whom the census question was asked, as reflected on the census questionnaire. For example, children are not usually asked employment questions, and men and children are not asked fertility questions. Cases that are outside of the universe for a variable are labeled "NIU" on the codes page. Differences in a variable's universe across samples are a common data comparability issue' (accessed at: https://cps.ipums.org/cps-action/faq#ques8). If percentages do not total 100%, either data is missing or not in universe.

Table 1.7 Crosstabulation table of DHS country/year and respondent's highest educational level in 20 countries on the African continent

	Highest educational level (percentage)				*Total (count)*
	None	*Primary*	*Secondary*	*Higher*	
Benin 1996	73.3	18.5	7.7	0.4	5,491
Benin 2001	64.9	21.6	12.6	0.8	6,219
Benin 2006	65.1	19.4	14.6	0.9	17,794
Benin 2011	62.6	16.7	19.4	1.4	16,599
Burkina Faso 1993	73.5	14.4	11.3	0.9	6,354
Burkina Faso 1998	81.3	10.0	8.3	0.4	6,445
Burkina Faso 2003	79.4	11.8	8.5	0.3	12,477
Burkina Faso 2010	73.0	14.1	12.2	0.7	17,087
Cameroon 1991	33.0	32.9	32.4	1.7	3,871
Cameroon 1998	24.2	36.2	37.4	2.2	5,501
Cameroon 2004	20.1	40.4	37.5	2.0	10,656
Cameroon 2011	18.1	35.5	41.7	4.7	15,426
Côte d'Ivoire 1994	60.6	25.1	13.7	0.6	8,099
Côte d'Ivoire 1998	49.5	29.3	18.7	2.4	3,040
Côte d'Ivoire 2011	57.1	23.3	17.3	2.3	10,060
Egypt 1988	49.7	32.4	13.0	4.9	8,911
Egypt 1992	47.0	26.0	22.3	4.6	9,864
Egypt 1995	46.0	23.9	24.6	5.5	14,779
Egypt 2000	42.5	18.2	31.3	8.0	15,573
Egypt 2005	35.6	15.7	39.4	9.3	19,474
Egypt 2008	33.5	12.4	43.2	10.9	16,527
Egypt 2014	22.3	10.0	53.0	14.7	21,762
Ethiopia 2000	68.9	16.5	13.6	1.0	15,367
Ethiopia 2005	60.1	21.1	16.3	2.5	14,070
Ethiopia 2011	50.1	35.5	8.4	6.0	16,515
Ghana 1988	39.7	52.8	6.6	0.9	4,488
Ghana 1993	35.0	54.7	8.7	1.6	4,562
Ghana 1998	35.9	16.8	45.2	2.2	4,843
Ghana 2003	33.7	19.5	44.2	2.5	5,691
Ghana 2008	25.3	20.3	50.7	3.7	4,916
Guinea 1999	79.4	10.7	7.8	2.1	6,753
Guinea 2005	78.3	10.9	10.3	0.6	7,954
Guinea 2012	67.4	13.9	15.5	3.2	9,142
Kenya 1989	23.8	53.5	22.2	0.3	7,150
Kenya 1993	17.2	59.0	23.3	0.5	7,540
Kenya 1998	12.8	59.9	25.4	1.9	7,881
Kenya 2003	15.8	53.1	24.1	7.1	8,195
Kenya 2008–2009	14.7	52.2	24.7	8.5	8,444
Madagascar 1992	16.8	49.1	30.9	3.1	6,260
Madagascar 1997	20.8	48.7	27.9	2.6	7,060
Madagascar 2003	15.4	39.6	40.3	4.8	7,949
Madagascar 2008	20.7	45.9	30.4	3.0	17,375

(*continued*)

Table 1.7 (cont.)

	Highest educational level (percentage)				*Total (count)*
	None	*Primary*	*Secondary*	*Higher*	
Malawi 1992	37.8	54.3	7.6	0.3	4,849
Malawi 2000	25.5	62.2	12.2	0.2	13,220
Malawi 2004	23.3	62.2	13.8	0.6	11,698
Malawi 2010	14.7	66.6	17.2	1.4	23,020
Mali 1987	80.4	17.8	1.7	0.2	3,200
Mali 1995–1996	80.1	12.6	7.1	0.3	9,704
Mali 2001	79.9	11.4	8.0	0.7	12,849
Mali 2006	77.7	11.7	10.0	0.6	14,583
Mali 2012	74.1	9.7	14.7	1.5	10,424
Mozambique 1997	39.1	55.2	5.5	0.2	8,779
Mozambique 2003	36.2	54.1	9.4	0.3	12,418
Mozambique 2011	27.4	49.3	21.4	1.9	13,745
Niger 1992	81.5	11.2	7.1	0.3	6,503
Niger 1998	80.1	12.3	7.2	0.4	7,577
Niger 2006	76.8	12.7	9.9	0.7	9,223
Niger 2012	74.6	12.9	11.3	1.0	11,160
Nigeria 1990	51.7	24.1	21.8	2.5	8,781
Nigeria 1999	37.8	27.4	29.5	5.3	9,810
Nigeria 2003	39.4	21.9	32.3	6.4	7,620
Nigeria 2008	39.7	19.7	32.7	7.9	33,385
Nigeria 2013	35.3	18.2	37.0	9.5	38,948
Rwanda 1992	35.8	53.3	10.6	0.4	6,551
Rwanda 2000	27.2	59.3	12.9	0.7	10,421
Rwanda 2005	23.0	66.2	10.0	0.8	11,321
Rwanda 2010	15.1	67.9	15.3	1.8	13,671
Tanzania 1991	35.2	60.0	4.6	0.2	9,238
Tanzania 1996	27.6	65.5	6.7	0.0	8,120
Tanzania 1999	25.5	61.1	13.4	0.0	4,029
Tanzania 2004	24.5	61.8	11.7	2.0	10,329
Tanzania 2010	18.9	57.9	22.8	0.5	10,139
Uganda 1988	34.5	52.7	12.5	0.3	4,730
Uganda 1995	25.6	55.2	18.9	0.4	7,070
Uganda 2001	20.1	56.6	18.7	4.6	7,246
Uganda 2006	20.7	57.7	17.5	4.0	8,531
Uganda 2011	15.4	55.6	22.7	6.3	8,674
Zambia 1992	17.2	60.1	21.0	1.6	7,060
Zambia 1996	14.6	60.3	22.8	2.4	8,021
Zambia 2001	13.1	59.2	25.0	2.7	7,658
Zambia 2007	10.4	53.2	31.4	5.0	7,146
Zambia 2013	8.3	46.6	39.9	5.2	16,411
Zimbabwe 1988	13.5	55.9	29.7	0.9	4,201
Zimbabwe 1994	11.6	48.3	38.8	1.3	6,128
Zimbabwe 1999	7.4	42.6	47.5	2.5	5,907
Zimbabwe 2005–2006	4.3	33.4	59.5	2.9	8,907

	Highest educational level (percentage)				*Total (count)*
	None	*Primary*	*Secondary*	*Higher*	
Zimbabwe 2010–2011	2.4	28.9	64.4	4.3	9,171
Zimbabwe 2015	1.3	25.8	65.6	7.3	9,955
Total count	353,342	305,821	206,355	30,750	896,340
Total percentage	39.4	34.1	23.0	3.4	100.0

Citation: Elizabeth Heger Boyle, Miriam King, and Matthew Sobek. IPUMS-Demographic and Health Surveys: Version 4.1 (dataset). Minnesota Population Center and ICF International, 2017. http://doi.org/10.18128/D080.V4.0.

Source: ICF: The DHS Model Datasets 2017 (dataset), Rockville, MD, USA: ICF. Data extract from all datasets noted above. IPUMS-Demographic and Health Surveys (IPUMSDHS), version 4.1, Minnesota Population Center and ICF International (Distributors). Accessed at: http://idhsdata.org

Associations with household cell phone ownership

Preliminary chi-square analysis allows me to explore factors associated with a cell phone in the households of childbearing-aged women captured in DHS's representative samples. From this analysis, we learn that women in households with cell phones have a more liberal attitude toward intimate partner violence, namely that it is not justified. Table 1.8 examines the crosstabulation of whether a household has a cell phone and attitudes that wife beating is justified if a woman refuses to have sex. Respondents (women, mostly wives of male household heads) in households that own cell phones are more likely to disagree with the idea that wife beating is justified if a woman refuses sex. A majority (77% of women whose households own cell phones) said wife beating is not justified (though 20% said it is justified). According to chi-square results, this association is highly statistically significant.

Additionally, the higher a woman's educational level, the more likely it is that someone in her household owns a cell phone. Table 1.9 examines the crosstabulation of whether a household has a cell phone and the highest educational level of female respondents. Households in which female respondents have secondary or further education are more likely to own cell phones, relative to households in which women respondents have either primary level or no formal education. The association between education of the childbearing-aged female respondents and cell phone ownership is highly statistically significant.

Women in households with cell phones have better access to health care. In examining the association between cell phone ownership and

Table 1.8 Crosstabulation table of whether household has cell phone by attitude to wife beating: justified if woman refuses to have sex (DHS data from 19 African countries)

		Attitude to wife beating: justified if woman refuses to have sex				
		No	*Yes*	*Don't know*	*Missing*	*Total*
Household has cell phone	No	14,688 68.7%	59,734 27.9%	7,012 3.3%	211 0.1%	21,3842 100%
	Yes	174,266 77.2%	46,026 20.4%	5,326 2.4%	157 0.1%	225,775 100%
	Missing	268 71.8%	85 22.8%	20 5.4%	0 0%	373 100%
Total		321,419 73.1%	105,845 24.1%	12,358 2.8%	368 0.1%	439,990 100%

Chi-square tests	*Value*	*df*	*Asymptotic significance (two-sided)*
Pearson chi-square	4,037.788*	6	0.0000
Likelihood ratio	4,041.968	6	0.0000
Linear-by-linear association	1,195.518	1	0.0000
No. of valid cases	439,990		

Citation: Elizabeth Heger Boyle, Miriam King, and Matthew Sobek. IPUMS-Demographic and Health Surveys: Version 4.1 (dataset). Minnesota Population Center and ICF International, 2017. http://doi.org/10.18128/D080.V4.0.

Source: ICF: The DHS Model Datasets 2017 (dataset), Rockville, MD, USA: ICF. Data extract from all datasets noted above. IPUMS-Demographic and Health Surveys (IPUMSDHS), version 4.1, Minnesota Population Center and ICF International (Distributors). Accessed at: http://idhsdata.org

* Cells (8.3%) have expected count of fewer than 5. The minimum expected count is .31.

Note: See Table 1.3 for list of 19 countries and years of data.

whether distance to medical facility is a barrier to woman obtaining their own health care, in more households in which women say distance is a "big problem" in terms of accessing health care, members do not own cell phones compared to households where women do not say distance is a big problem. This difference is highly statistically significant. See Table 1.10.

Table 1.9 Crosstabulation table of female respondent's highest educational level by whether her household has a cell phone (DHS data from 19 African countries)

			Highest educational level				
			No education	*Primary*	*Secondary*	*Higher*	*Missing*
Household has cell phone	No		94,608	80,778	36,083	2,365	8
			44.2%	37.8%	16.9%	1.1%	0%
	Yes		59,548	59,954	87,809	18,432	32
			26.4%	26.6%	38.9%	8.2%	0%
	Missing		122	117	115	19	0
			32.7%	31.4%	30.8%	5.1%	0%
Total			154,278	140,849	124,007	20,816	40
			35.1%	32%	28.2%	4.7%	0%
Chi-square tests	*Value*	*df*	*Asymptotic significance (two-sided)*				
Pearson chi-square	44,787.502*	8	0.0000				
Likelihood ratio	47,198.984	8	0.0000				
Linear-by-linear association	32,860.035	1	0.0000				
No. of valid cases	439,990						

Citation: Elizabeth Heger Boyle, Miriam King, and Matthew Sobek. IPUMS-Demographic and Health Surveys: Version 4.1 (dataset). Minnesota Population Center and ICF International, 2017. http://doi.org/10.18128/D080.V4.0.

Source: ICF: The DHS Model Datasets 2017 (dataset), Rockville, MD, USA: ICF. Data extract from all datasets noted above. IPUMS- Demographic and Health Surveys (IPUMSDHS), version 4.1, Minnesota Population Center and ICF International (Distributors). Accessed at: http://idhsdata.org

* One cell (6.7%) has expected count of fewer than 5. The minimum expected count is .03.

Note: See Table 1.3 for the list of 19 countries and years of data collection.

Table 1.10 Crosstabulation table of whether a household has a cell phone and distance as a barrier to female respondent's health care access (DHS data from 19 African countries)

		Barrier to woman's health care: distance to facility				
		Not a big problem	*Is a big problem*	*Missing*	*NIU (not in universe)**	*Total*
Household has cell phone	No	62,023	43,264	100	5,968	111,355
		55.7%	38.9%	0.1%	5.4%	100%
	Yes	78,033	31,900	86	2,849	112,868
		69.1%	28.3%	0.1%	2.5%	100%
	Missing	94	69	1	4	168
		56%	41.1%	0.6%	2.4%	100%
Total		140,150	75,233	187	8,821	224,391
		62.5%	33.5%	0.1%	3.9%	100%

Chi-square tests	*Value*	*df*	*Asymptotic significance (two-sided)*
Pearson chi-square	4,037.788a	6	0.0000
Likelihood ratio	4,041.968	6	0.0000
Linear-by-linear association	1,195.518	1	0.0000
No. of valid cases	439,990		

Citation: Elizabeth Heger Boyle, Miriam King, and Matthew Sobek. IPUMS-Demographic and Health Surveys: Version 4.1 (dataset). Minnesota Population Center and ICF International, 2017. http://doi.org/10.18128/D080.V4.0.

Source: ICF: The DHS Model Datasets 2017 (dataset), Rockville, MD, USA: ICF. Data extract from all datasets noted above. IPUMS- Demographic and Health Surveys (IPUMSDHS), version 4.1, Minnesota Population Center and ICF International (Distributors). Accessed at: http://idhsdata.org

* According to the IDHS, 'The universe is the population at risk of having a response for the variable in question. In most cases these are the households or persons to whom the census question was asked, as reflected on the census questionnaire. For example, children are not usually asked employment questions, and men and children are not asked fertility questions. Cases that are outside of the universe for a variable are labeled "NIU" on the codes page. Differences in a variable's universe across samples are a common data comparability issue' (accessed at: https://cps.ipums.org/cps-action/faq#ques8).

Notes:
See Table 1.3 for the list of 19 countries and years of data collection.
Two cells (8.3%) have expected count of fewer than 5. The minimum expected count is .31.

Table 1.11 Crosstabulation table of respondent's relationship to household head and whether household has a cell phone (DHS data from 19 African countries)

		Relationship to household head			
		Head	*Wife*	*Other*	*Total*
Household has cell phone	No	28,731	125,945	59,166	213,842
		13.4%	58.9%	27.7%	100%
	Yes	21,606	127,360	76,809	225,775
		9.6%	56.4%	34%	100%
	Missing	31	216	126	373
		8.3%	57.9%	33.8%	100%
Total		50,368	253,521	136,101	439,990
		11.4%	57.6%	30.9%	100%

Chi-square tests	*Value*	*df*	*Asymptotic significance (two-sided)*
Pearson chi-square	2988.332*	4	0.0000
Likelihood ratio	2996.047	4	0.0000
Linear-by-linear association	2571.829	1	0.0000
No. of valid cases	439,990		

Citation: Elizabeth Heger Boyle, Miriam King, and Matthew Sobek. IPUMS-Demographic and Health Surveys: Version 4.1 (dataset). Minnesota Population Center and ICF International, 2017. http://doi.org/10.18128/D080.V4.0.

Source: ICF: The DHS Model Datasets 2017 (dataset), Rockville, MD, USA: ICF. Data extract from all datasets noted above. IPUMS-Demographic and Health Surveys (IPUMSDHS), version 4.1, Minnesota Population Center and ICF International (Distributors). Accessed at: http://idhsdata.org

* Cells (.0%) have expected count of fewer than 5. The minimum expected count is 42.70.

Note: See Table 1.3 for the list of 19 countries and years of data collection.

Households with male heads are more likely to have a member who owns a cell phone. Sex of the household head and the relationship to the household head are examined in Tables 1.11 and 1.12. If a female respondent is the household head, she is less likely to own a cell phone compared to women in households who are wives of the household head. And if the household head is female, her household is less likely to own a cell phone compared to women in households with a male head. Both differences are statistically significant.

Married women are more likely to live in homes in which a member owns a cell phone. The (all-female) respondents' current marital or

Table 1.12 Crosstabulation table of sex of household head and whether household has a cell phone (DHS data from 19 African countries)

		Sex of household head		
		Male	*Female*	*Total*
Household has cell phone	No	163,359	50,483	213,842
		76.4%	23.6%	100%
	Yes	181,305	44,470	225,775
		80.3%	19.7%	100%
	Missing	300	73	373
		80.4%	19.6%	100%
Total		344,964	95,026	439,990
		78.4%	21.6%	100%

Chi-square tests	*Value*	*df*	*Asymptotic significance (two-sided)*
Pearson chi-square	992.977*	2	0.0000
Likelihood ratio	992.848	2	0.0000
Linear-by-linear association	855.642	1	0.0000
No. of valid cases	439,990		

Citation: Elizabeth Heger Boyle, Miriam King, and Matthew Sobek. IPUMS-Demographic and Health Surveys.

Source: ICF: The DHS Model Datasets 2017 (dataset), Rockville, MD, USA: ICF. Data extract from all datasets noted above. IPUMS-Demographic and Health Surveys (IPUMSDHS), version 4.1, Minnesota Population Center and ICF International (Distributors). Accessed at: http://idhsdata.org

* Cells (.0%) have expected count of fewer than 5. The minimum expected count is 80.56.

Note: See Table 1.3 for the list of 19 countries and years of data collection.

union status and its association with household cell phone ownership is examined in Table 1.13. Women who have never married are more likely to be in households in which a cell phone is owned (24% versus 17%). Currently married women are slightly less likely to be in households in which a cell phone is owned. Women who are widowed, divorced, or separated are less likely to be in households that own a cell phone.

And finally, younger women are more likely to live in homes in which a member owns a cell phone. Age differences are examined in Table 1.14. With the backdrop of associations between marital status and cell phone ownership, the age data follow a familiar pattern. Among female respondents aged 40 and over (up to 49), less than half of respondents are in households in which cell phones are owned.

Table 1.13 Crosstabulation table of female respondent's marital status and whether her household has a cell phone (DHS data from 19 African countries)

		Woman's current marital or union status							
		Never married	*Married*	*Living together*	*Widowed*	*Divorced*	*Separated/not living together*	*Missing*	*Total*
Household has cell phone	No	35,919	142,592	13,639	8,273	6,080	7,337	2	213,842
		16.8%	66.7%	6.4%	3.9%	2.8%	3.4%	0%	100%
	Yes	55,014	140,159	15,546	5,575	4,122	5,354	5	225,775
		24.4%	62.1%	6.9%	2.5%	1.8%	2.4%	0%	100%
	Missing	79	244	18	14	5	13	0	373
		21.2%	65.4%	4.8%	3.8%	1.3%	3.5%	0%	100%
Total		91,012	282,995	29,203	13,862	10,207	12,704	7	439,990
		20.7%	64.3%	6.6%	3.2%	2.3%	2.9%	0%	100%

Chi-square tests	*Value*	*df*	*Asymptotic significance (two-sided)*
Pearson chi-square	5052.011*	12	0.0000
Likelihood ratio	5085.869	12	0.0000
Linear-by-linear association	3995.657	1	0.0000
No. of valid cases	439,990		

Citation: Elizabeth Heger Boyle, Miriam King, and Matthew Sobek. IPUMS-Demographic and Health Surveys: Version 4.1 (dataset). Minnesota Population Center and ICF International, 2017. http://doi.org/10.18128/D080.V4.0.

Source: ICF: The DHS Model Datasets 2017 (dataset), Rockville, MD, USA: ICF. Data extract from all datasets noted above. IPUMS- Demographic and Health Surveys (IPUMSDHS), version 4.1, Minnesota Population Center and ICF International (Distributors). Accessed at: http://idhsdata.org

* Three cells (14.3%) have expected count of fewer than 5. The minimum expected count is .01.

Note: See Table 1.3 for the list of 19 countries and years of data collection.

Table 1.14 Crosstabulation table of respondent's age by whether the household has a cell phone (DHS data from 19 African countries)

Age	*Yes*	*Missing*	*Total*
15	8,986	9	17,678
	50.8%	0.1%	100%
16	8,517	16	16,638
	51.2%	0.1%	100%
17	8,195	13	15,565
	52.7%	0.1%	100%
18	9,249	21	18,218
	50.8%	0.1%	100%
19	7,801	16	14,563
	53.6%	0.1%	100%
20	10,412	23	20,854
	49.9%	0.1%	100%
21	7,353	7	13,488
	54.5%	0.1%	100%
22	8,937	14	16,968
	52.7%	0.1%	100%
23	7,928	12	14,741
	53.8%	0.1%	100%
24	7,777	14	14,471
	53.7%	0.1%	100%
25	10952	14	21,934
	49.9%	0.1%	100%
26	8,204	16	14,938
	54.9%	0.1%	100%
27	7,995	13	14,836
	53.9%	0.1%	100%
28	8,922	13	16,746
	53.3%	0.1%	100%
29	6,705	10	12,042
	55.7%	0.1%	100%
30	10,061	21	20,814
	48.3%	0.1%	100%

Age	*Yes*	*Missing*	*Total*
31	5,567	4	10,243
	54.3%	0%	100%
32	6,631	12	12,867
	51.5%	0.1%	100%
33	5,506	7	10,179
	54.1%	0.1%	100%
34	5,269	2	9,813
	53.7%	0%	100%
35	7,771	5	16,294
	47.7%	0%	100%
36	4,998	12	9,775
	51.1%	0.1%	100%
37	4,772	5	9,204
	51.8%	0.1%	100%
38	5,217	10	10,418
	50.1%	0.1%	100%
39	4,079	6	7,781
	52.4%	0.1%	100%
40	6,443	20	14,177
	45.4%	0.1%	100%
41	3,477	4	6,651
	52.3%	0.1%	100%
42	4,100	9	8,294
	49.4%	0.1%	100%
43	3,477	5	6,948
	50%	0.1%	100%
44	3,127	4	6,050
	51.7%	0.1%	100%
45	4,739	10	10,418
	45.5%	0.1%	100%
46	3,030	6	6,406
	47.3%	0.1%	100%
47	3,064	5	6,239
	49.1%	0.1%	100%

(*continued*)

Table 1.14 (cont.)

Age	*Yes*	*Missing*	*Total*
48	3,502	6	7,449
	47%	0.1%	100%
49	3,012	9	6,290
	47.9%	0.1%	100%
Total count	225,775	373	439,990
Total percentage	51.3%	0.1%	100%

Chi-square tests	*Value*	*df*	*Asymptotic significance (two-sided)*
Pearson chi-square	1242.377*	68	0.0000
Likelihood ratio	1246.765	68	0.0000
Linear-by-linear association	216.541	1	0.0000
No. of valid cases	439,990		

Citation: Elizabeth Heger Boyle, Miriam King, and Matthew Sobek. IPUMS-Demographic and Health Surveys: Version 4.1 (dataset). Minnesota Population Center and ICF International, 2017. http://doi.org/10.18128/D080.V4.0.

Source: ICF: The DHS Model Datasets 2017 (dataset), Rockville, MD, USA: ICF. Data extract from all datasets noted above. IPUMS- Demographic and Health Surveys (IPUMSDHS), version 4.1, Minnesota Population Center and ICF International (Distributors). Accessed at: http://idhsdata.org

* Cells (.0%) have expected count of fewer than 5.

Note: See Table 1.3 for the list of 19 countries and years of data collection.

Among female respondents aged 30–39, more than half of respondents are in households in which cell phones are owned for most ages. Among female respondents aged 15–29, more than half are in households in which cell phones are owned. So, the youngest respondents (aged 15–29) are more likely to reside in households that own cell phones compared to all older women. And the oldest women (aged 40–49) are less likely to be in households in which cell phones are owned. These differences are statistically significant.

Summary: marginalized users and the gender digital divide

In examining aggregate data from African nations, mobile-cellular telephone subscriptions overall and account ownership among women are

Table 1.15 Cell phones in women's households and growth in cell phone markets since last DHS data collection (includes countries for analysis of impact on inclusion, women's education, and governance)

Country and year	*Cell phones in women's households?*	*Rate of increase from time of last DHS*
Egypt 2005	22.6%	NA
Egypt 2008	40.7%	18.1
Egypt 2014	92.1%	51.4
Ethiopia 2005	7.3%	NA
Kenya 2008–2009	64.4%	NA
Kenya 2014	86.0%	22
Malawi 2004	5.5%	NA
Malawi 2010	43.2%	37.7
Mozambique 2011	52.9%	NA
Nigeria 2008	48.0%	NA
Nigeria 2013	78.5%	30.5
Nigeria 2015	78.5%	0
Tanzania 2010	56.6%	NA
Tanzania 2015	78.0%	21
Uganda 2006	20.3%	NA
Uganda 2011	65.6%	45.2
Uganda 2014–2015	67.9%	2
Zambia 2007	38.3%	NA
Zambia 2013	73.6%	35.3
Zimbabwe 2005–2006	17.1%	NA
Zimbabwe 2010–2011	66.6%	49.50
Zimbabwe 2015	87.0%	20
Average	51.0%	NA

Source: Zerai's analysis of data extract performed by DHS implementing partners and ICF International. Demographic and Health Surveys (see table for sources). Data extract from DHS Recode files. IPUMS-Demographic and Health Surveys (IPUMS-DHS), version 3.0, Minnesota Population Center and ICF International (Distributors). Accessed at: http://idhsdata.org

Note: Data not available in IPUMS at the time this table was created was added from Demographic and Health Surveys from Kenya 2014, Nigeria 2015, Tanzania 2015; Uganda 2014–2015; Zimbabwe 2015 (see bibliography for citations).

increasing. Table 1.15 provides data on the increase overtime in the numbers of cell phones in the households of childbearing-aged women. These data are available from the DHS. All countries with available data on cell phones in households demonstrate an increase compared to the last survey, with the exception of Nigeria, which does not show an increase from 2013 to 2015; however, the number of cell phones there remains high at 78.5%.

Internet access is also increasing; however, the gender breakdown of these data is not available. In household data from 19 countries in the DHS, descriptive data provide a profile of women in households that own cell phones. I find that phone ownership is associated with women who are youthful, highly educated, residing in urban areas, with higher socioeconomic status, and among groups that are less likely to think that intimate partner violence is acceptable.

In summary, female respondents who are household heads are less likely to own a cell phone compared to women who are wives of household heads. Women who are widowed, divorced, or separated are less likely to be in households that own a cell phone. Women who indicate that access to health facilities is a big problem are in households that are less likely to have cell phones. And women who agree that intimate partner violence is acceptable if a woman refuses to have sex are less likely to live in households with a cell phone. Thus, women who need phones the most do not have them.

Bivariate analyses and reexamination of Asongu and Nwachukwu's models

Moving beyond a description of households in which there is a cell phone in 19 countries, I next examine three focus countries that provide information on *women's* cell phone ownership. Into an aggregate file I join Zimbabwe (2010, 2015), Tanzania (2010, 2016), and Malawi (2010, 2015) data from the DHS, Worldwide Governance Indicators (WGI), and Country Policy and Institutional Assessment (CPIA) for these countries/years into an analysis file (see Table 1.16 for sample sizes in the six DHS datasets). In the new three-country file, from chi-square results, I find that an average of 64% of households owned cell phones (in 2010 and 2015/2016) and 47% of respondents (women aged 15–49) owned cell phones in 2015 or 2016: ranging from 35% in Malawi, 52% in Tanzania, to 72% in Zimbabwe (see Table 1.17).

Conceptual models: from Asongu and Nwachukwu's androcentric knowledge diffusion hypothesis to considering gender-sensitive governance indices

Asongu and Nwachukwu's (2016) analysis of 2000–2012 World Bank development and governance indicators, provides evidence that cell

Table 1.16 Data utilized for three-country analyses: number of respondents (women aged 15–49)

	Frequency	*Percentage*	*Cumulative percentage*
Malawi 2010	23,020	25.5	25.5
Malawi 2016	24,562	27.3	52.8
Tanzania 2010	10,139	11.3	85.3
Tanzania 2015	13,266	14.7	100
Zimbabwe 2010–2011	9,171	10.2	63
Zimbabwe 2015	9,955	11	74
Total	90,113	100	

Source: Zerai's analysis of data extract performed by DHS implementing partners and ICF International. Demographic and Health Surveys Malawi 2010, 2016; Tanzania 2010, 2015–2016; Zimbabwe 2010, 2015. Data extract from DHS Recode files. IPUMS-Demographic and Health Surveys (IPUMS-DHS), version 3.0, Minnesota Population Center and ICF International (Distributors). Accessed at: http://idhsdata.org

Table 1.17 Percentage of women aged 15–49 who own a cell phone in latest Malawi, Tanzania, and Zimbabwe DHS

	Female respondent does not own cell phone	*Female respondent owns cell phone*	*Total female respondents*	*Percentage of women 15–49 who own a cell phone*
Malawi 2016	15,990	8,572	24,562	34.90%
Tanzania 2015	6,428	6,838	13,266	51.55%
Zimbabwe 2015	2,785	7,170	9,955	72.02%
Total	25,203	22,580	47,783	47.26%

Source: Zerai's analysis of data extract performed by DHS implementing partners and ICF International. Demographic and Health Surveys Malawi 2010, 2016; Tanzania 2010, 2015–2016; Zimbabwe 2010, 2015–2016. Data extract from DHS Recode files. IPUMS-Demographic and Health Surveys (IPUMS-DHS), version 3.0, Minnesota Population Center and ICF International [Distributors]. Accessed at: http://idhsdata.org

phones/ICT are linked to governance indicators such as: the rule of law, corruption-control, regulation quality, government effectiveness, political stability/no violence and voice and accountability. Their work shows that diffusion of knowledge is an important mechanism linking ICT and better governance. Below, I create a conceptual model of their argument and delineate how we could operationalize their theory utilizing DHS data. In my quest to challenge both their definitions of good governance as well as their conclusion that implies that a gender-blind understanding of diffusion of knowledge sufficiently explains factors resulting in better governance in Sub-Saharan Africa, I begin with

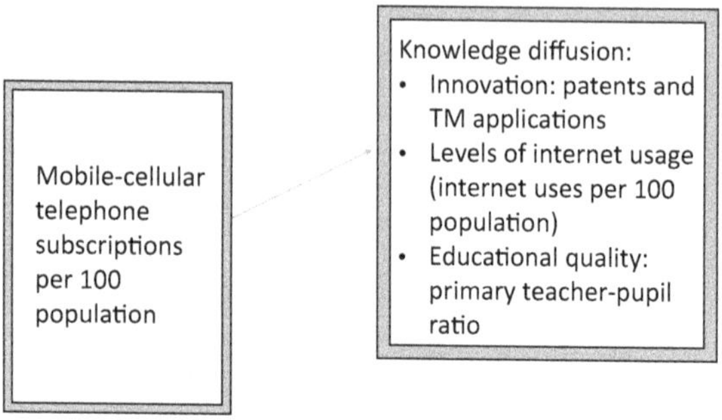

Figure 1.1 Conceptualizing the ICT and knowledge diffusion hypothesis in Asongu and Nwachukwu's model (2016)

replicating some of Asongu and Nwachukwu's findings in relation to the knowledge economy.[1] Figure 1.1 provides a basic summary of Asongu and Nwachukwu's theory. First, it indicates that ICT facilitates knowledge diffusion, in the form of education/educational quality, internet use, and also in terms of innovation. Focusing on educational quality, I therefore examine the extent to which cell phones in households are associated with knowledge diffusion amongst women in Zimbabwe, Tanzania, and Malawi (Figure 1.2). These three countries are selected because their DHS uniquely boasts data on women's indivudal ownership of cell phones in 2015–2016. The distinction in my analysis is that instead of running a general method of moments analysis of 54 countries, where countries are the unit of analysis (as in the case of Asongu and Nwachukwu (2016), I make the households of women the unit of analysis. My hypothesis is that if I reproduce an analysis on the basis of their theory, ICT (in the form of cell phones in women's households) will be associated with knowledge diffusion (as measured by literacy and media consumption), as evidenced by statistically significant results in the bivariate case. I ran basic chi-square analyses to test this hypothesis. I found that a cell phone in a woman's household is associated with her literacy,[2] her highest educational level, and her degree of media consumption.[3]

Further, I examined the extent to which cell phones are associated with country-level indicators of knowledge diffusion (Figure 1.3). Looking at cell phones in the home and World Bank development indicators, cell phones are associated with: higher primary level completion rates (countrywide), lower pupil-teacher ratios (as in Asongu and Nwachukwu (2016), a school enrollment gender parity index of 1.0, and higher percentages of individuals using the internet. Next, I examined

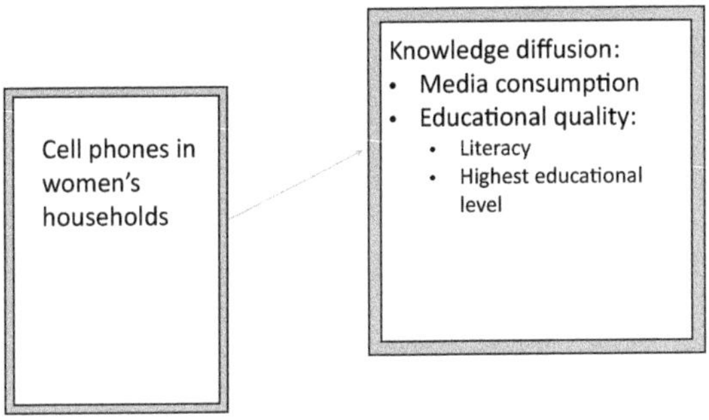

Figure 1.2 Examining an ICT and knowledge diffusion model with DHS data (women aged 15–49)

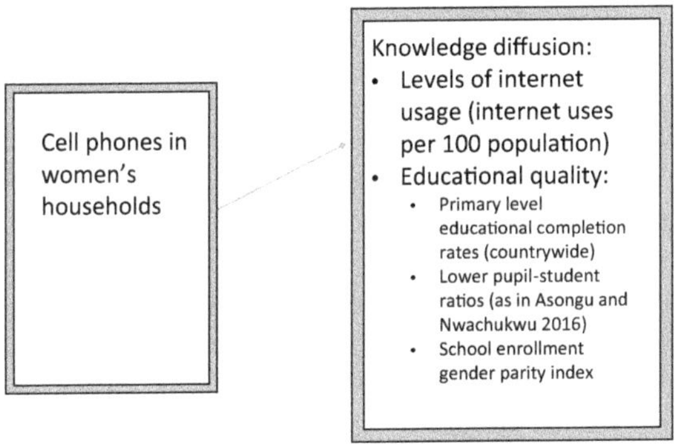

Figure 1.3 Examining an ICT and knowledge diffusion model with DHS and World Bank data (among three countries)

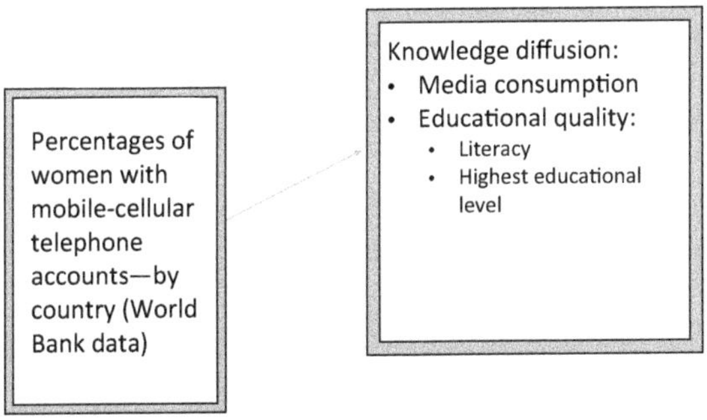

Figure 1.4 ICT and knowledge diffusion hypothesis with women's mobile-cellular telephone accounts

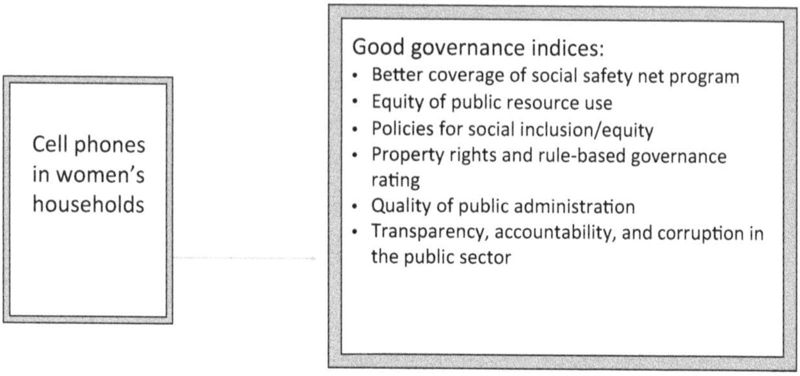

Figure 1.5 Cell phones in women's households and non-gender-specific governance indices

the percentages of women with mobile accounts to see what impact this had on knowledge diffusion (Figure 1.4). Lo and behold, countries with higher percentages of women with cell phone accounts (from World Bank data) have higher completed educational levels, higher literacy, and higher levels of media consumption.[4]

Finally, I examined bivariate associations between cell phones in women's households and WGI and CPIA indices in Malawi 2016, Tanzania 2015–2016, and Zimbabwe 2015–2016 (Figure 1.5).

To summarize: in relation to *nongender-specific governance ratings* from WGI and CPIA, cell phones in women's households are related to:

- better coverage of the social safety net program;
- equity of public resource use;
- policies for social inclusion/equity;
- property rights and rule-based governance rating;
- quality of the public administration;
- transparency, accountability, and corruption in the public sector.

Examining women-centered variables, cell phones in women's households is a variable that is related to a number of gender equality factors (Figure 1.6), which include:

- the law mandates equal remuneration for females and males for work of equal value;
- legislation exists on domestic violence;

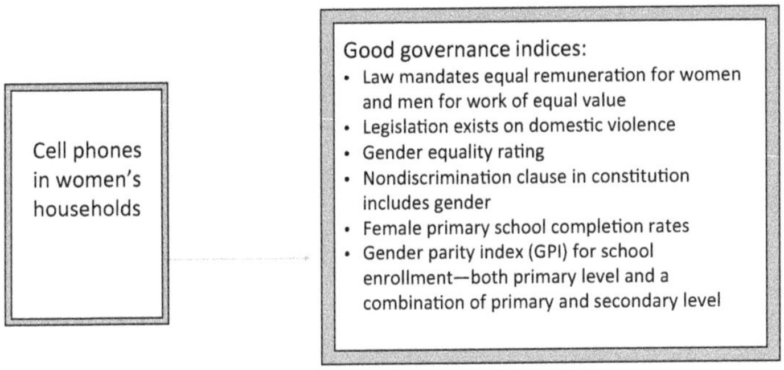

Figure 1.6 Cell phones in women's households and gender-sensitive governance indices

- gender equality rating;
- a nondiscrimination clause in constitution includes gender;
- female primary school completion rates;
- gender parity index (GPI) for school enrollment—both primary level and a combination of primary and secondary levels.

Chi-square results in the bivariate case are significant at $p<.001$ for all associations noted above, meaning that the two variables are not independent (results are not shown).[5]

Building on the analyses here, in Chapters 2 to 4, I utilize a female scholars' database and examine DHS data, World Bank WDI, WGI, and CPIA development and governance indicators in Zimbabwe, Tanzania, and Malawi in order to build a multivariate model to test for predictors of both the availability of mobile-cellular telephone technology in women's households and outcome factors that should result from government efforts to promote gender equity. Finally, in Chapter 5, I consider the literature of women scholars in Zimbabwe, Tanzania, Malawi, and from elsewhere in Africa and the African Diaspora to rethink a conceptual model to examine the impact of ICT on national governance factors. I focus on issues of gender equity identified in the review of indigenous scholars' literature and activist discourse.

Notes

1 In this brief analysis that validates some of Asongu and Nwachukwu's results, I examine these predictors separately. In subsequent chapters, I conduct multivariate logistic regression analyses.

2 The literacy variable is operationalized utilizing an ordinal scale in the following way. A women is literate if she has either completed secondary education or reads a passage easily. If she has not completed secondary education and cannot read a passage at all, she is not considered to be literate. If she has not completed secondary education and reads a passage with difficulty, she is considered partially literate.

3 Media consumption includes reads newspaper, watches television, or listens to the radio as a dichotomous variable.

4 For the sake of simplicity, chi-square results are not shown.

5 I also examined the proportion of women in the national parliament, but the results were not statistically significant.

2 ICT, women's status, and governance in Zimbabwe[1]

In this chapter, I start with women academics in Zimbabwean institutions to determine organic definitions of women's status, diffusion of knowledge, and quality governance that are relevant to women there. In discussing how I built the database for this chapter, I make the case for and explain details of my women-centered methodology. Then, I examine cell phone use and its impact on knowledge diffusion among women by outlining features of a conceptual model for analyzing the Zimbabwe DHS for 2005–2015 that have been merged with World Bank development and governance indicators. Further, I provide an analytic plan for examining the impact of ICT on community-level and national governance factors and values that focus on issues of gender equity identified in the review of indigenous scholars' literature and activist discourse. Zimbabwe is a fascinating context in which to study these issues. Within the country, cell phone ownership within households has increased from 17% in 2005–2006 to 67% in 2010–2011 (ZDHS 2012), and 87% in 2015 (ZDHS 2016), this despite authoritarian aspects of state governance. Remarkably, during the 2013 presidential election, one that occurred at a time when members of women's households were more likely to own a cell phone than at any other point previously, physical confrontations were benign compared to the pre-election violence surrounding presidential elections in previous decades (1990–2008). And of course, most recently, President Robert Mugabe was ousted by the military. The country hopes to return to democratic rule following general elections in July 2018, during which time citizens are expected to elect their president and seats in both Houses of Parliament. It was fascinating to watch the role of ICT in the election campaigns.

African feminism

Inspired by transnational feminist praxis (Dillard and Okpalaoka 2011), there is an exciting body of literature in the field that is broadly characterized as African feminism that has identified a perspective, a gendered lens, for myriad purposes (Gaidzanwa 1997; Mama 2002; Zerai 2014). The first and most important reason to articulate this woman-centered perspective has been to create a safe space for women to occupy on their own terms and in the context of male-dominated social structures (Imam, Mama, and Sow 1997). It has helped to legitimate the languages, discourses, challenges, unique perspectives, divergent experiences, and intersecting oppressions and privileges of African women's and girls' lives. I draw upon my (2014) emerging African feminist *methodology* to guide an exploratory discussion of women's scholarship and activism in Zimbabwe. My central argument is that in order to understand good governance, we must examine the impact of economic sanctions, state violence, infrastructure challenges, and access to ICT on gender equity. African women have worked to provide research and recommendations through their scholarship and activism. Yet, in our quest to understand how ICT and knowledge diffusion impact good governance, measures of good governance that are especially relevant to women have been ignored.

Western researchers often do not incorporate the voices of African women in their scholarly endeavors; and a serious exploration of dimensions of good governance in Zimbabwe cannot happen without this preliminary step. Building on the work of Franz Fanon (1963) and Lewis Gordon (2006), Reiland Rabaka (2010) refers to this as *epistemic apartheid*, "a process of critical decay within a field or discipline" due to its lack of intellectual diversity. This includes "institutional racism, academic colonization and conceptual quarantining of knowledge, anti-imperial thought, and/or radical political praxis produced and presented by … 'especially black' intellectual-activists" (Rabaka 2010, p. 16). To Rabaka's description, I add "especially" African *women* intellectual-activists. I ask how we engage with the scholarship and activism of African women to ensure that their voices and perspectives inform research on ICT and social development, especially when this research is focused on African contexts. In 2014 and 2016, I proposed an African feminist approach to promote intellectual diversity encompassing perspectives from Africa and the African Diaspora, that is consistent with Denzin and Lincoln's (2008) work on critical indigenous pedagogy (CIP).

> CIP is a productive dialogue between indigenous and critical scholars. This involves a re-visioning of critical pedagogy, a re-grounding of Paulo Freire's (2000) pedagogy of the oppressed in local, indigenous contexts. CIP understands that all inquiry is both political and moral. It uses methods critically for explicit social justice purposes. It values the transformative power of indigenous, subjugated knowledges. It values the pedagogical practices that produce these knowledges (Semali and Kincheloe 1999, p. 15), and it seeks forms of praxis and inquiry that are emancipatory and empowering. It embraces the commitment by indigenous scholars to decolonize Western methodologies, to criticize and demystify the ways in which Western science and the modern academy have been part of the colonial apparatus. This revisioning of critical pedagogy aligns with Paulo Freire and Antonio Faundez's (1989, p. 46) assertion that indigenous knowledge is a rich social resource for any justice-related attempt to bring about social change.
>
> (Denzin and Lincoln 2008, p. 2).

Pursuing African feminism provides the opportunity to create intellectual tools to build toward social justice instead of reproducing master narratives (cf. Audre Lorde). But such tools cannot be built while we remain in epistemic siloes. Rabaka (2010) reminds us to develop methodology that eschews epistemic apartheid and highlights the importance of foregrounding scholars and activists from the communities that we study (Zerai 2014). Unfortunately, many mainstream scholars still refuse to acknowledge the work of indigenous scholars who toil often without access to a wealth of resources. So, here searches of the work of scholars on the ground are prioritized. Some prominent African/Africana feminist scholars in (and those who have worked in) Zimbabwe are Rudo Gaidzanwa (1991, 1992, 1993, 1994, 1996, 1997, 1998, 1999a, 1999b, 2006, 2008, 2011, 2013), Patricia McFadden (1992, 1994, 2005, 2007a, 2007b, 2008, 2010), Makini Roy-Campbell (2001), Tanya Lyons (2004), Merle Bowen and Ayesha Tillman (2014), and Terri Barnes (Barnes 1995, 1999, 2002; Barnes and Win 1992), and the late Gloria Waite (2000). Gender equity-conscious male scholars include Horace Campbell (2003), Rob Davies (2004), and Michael West (1994, 2002), and the late Sam Moyo (Moyo 1995, 2000, 2007, 2008; Moyo, Makumbe and Raftopoulos 2000; Moyo and Yeros 2001; and Moyo and Amanor 2008). I cannot write about ICT and good governance in Zimbabwe without honoring their perspectives. The works of these scholars and other notable progressive African feminists, such as

Amina Mama (1995, 2002), Aisha Imam (1997), Patricia Daley (2008, 2014), and others, articulate the ultimate goal to affect social, political, and economic conditions for the purpose of enhancing the wellbeing of all African families.[2]

These works describe an African feminism that is consistent in some ways with Black feminist frameworks. Both recognize inequality/oppression as multiplicative and simultaneous (Zerai 2000b). However, some of the principal axes of oppression in Africa include gender, class, ethnicity, neocolonial relations, and heterosexual discourses/heterosexism, according to Imam (1997), Imam and Mama (1994), Imam, Mama, and Sow (1997), and others. Gender analysis highlights the necessity of considering ideology, subjectivity, and consciousness and the role of these "nonmaterial" processes in politics, production relations, democratic processes, and the state. Chikwenye Ogunyemi (1996) mentions corrupt governments, in-lawism, and fundamentalist religion as distinct experiences among African women. According to Mama (1995, 2002), Imam (1997), and others, multinational corporations, the International Monetary Fund (IMF) and the World Bank, US/EU commerce, sanctions and aid affect women on the continent of Africa differently from how they affect women of the African Diaspora.

Influences of the African Diaspora

In addition to crucial works by women of African descent in Africa, women in the African Diaspora express perspectives on gender that are instructive for African feminism. Adding to these theorists, I include aspects of the Black feminist framework elaborated elsewhere (and drawing on Brewer 1992, 1993, 1999; Collins 1990, 1998a, 1998b; Dillard 2000, 2011) (Zerai and Banks 2002), which are consistent with an emerging African feminism. Just as African feminism is a political project, Black feminist thought is a paradigm for reconceptualizing oppression and resistance (Collins 1990). The intersectionality of race, class, and gender is a framework arising from Black feminist thought for analyzing ways that various spheres of inequality operate simultaneously to affect social life.

With tenets emanating from African feminism and feminisms of the African Diaspora, including Black feminist theory and methods, I identify general principles for accomplishing an *African feminist methodology* (AFM) that may be applied to both political practice and research method. African feminisms and feminisms of the African Diaspora, including Black feminist thought, provide the basis for a more pan-African, or Africana, feminism. From the corpus of these theoretical works and political practice, we may begin to articulate an emerging African feminist methodology as a tool for praxis.

To summarize, AFM has three main features. This mixed methodology takes into account context and relationality, and produces an intersectional analysis. AFM takes *context* into account, some examples include that the social/political/economic context, historical antecedents, multiplicative social locations, and patriarchy/masculinity; globalization; state oppression and other inequalities are examined structurally. AFM examines *relationality*, including identifying how resources accrued to privileged groups directly relate to the withdrawal of resources from oppressed groups. Lastly, AFM produces *intersectional analysis* by examining ethnicity, class, gender, urbanity, region, and political party affiliation intersections, identifying these categories as multiplicative and simultaneous, and noting intersectional oppressions as well as intersections of resistance (for details, see Zerai and Banks 2002; and Zerai 2014). The focus of this manuscript is to highlight ways in which resources from women's ethnicity, gender, class, and neocolonial status *have not prevented their agency*, but have in fact given rise to their activism and scholarship. I examine a sample of this scholarship and activism in Zimbabwe, and also describe how I am building an African feminist scholar/activist database.

Database building methodology

Driven by my desire to understand Zimbabwe female academics' own perspectives on ICT, I captured often overlooked works of scholars in Zimbabwe by making a list of 15 universities in Zimbabwe and searching in all relevant departments for women faculty. Then I narrowed my search to the following units for the purposes of this chapter: mathematics and computer science, library and information science, social sciences (including sociology and social anthropology), e-commerce, communication skills, English and communication, gender and women's studies, development studies, politics and public management, media studies, institutes of peace, leadership and governance, management of administration, records and archives management, local governance studies, and statistics and operations research, in order to find scholars' curricula vitae. Next, I carefully read the various curricula vitae to search for a sampling of the articles relevant to the study of ICT in Zimbabwe.[3] From this, I generated a list of 63 female faculty members.[4]

Concerns of female and feminist/gender-conscious scholars in Zimbabwe

Numerous female African scholars in Zimbabwe work as faculty members in university settings and as active researchers. Their research fields that are relevant in this examination of ICT within the field of science and technology in society include computer science, library and information science, and records and archives management. Women's scholarship that is relevant to political, economic, and institutional governance includes fields such as social sciences, statistics and operations research, development studies, local governance studies, politics and public management (some authors include Ndiweni and Finch 1995, 1996; Wekwete 2014; and Chipfuwa et al. 2014). Female scholars' work that intersects with the study of knowledge diffusion can be found in fields such as education, e-commerce, communication skills, English and communication, and media studies. And their work that is relevant to gender equity can be found in fields such as gender and women's studies (Viriri et al. 2011), and other fields, as noted above.

In their recent scholarship, female scholars' research examines the broad areas of political governance, including women's empowerment, economic governance, knowledge diffusion, and ICT. Work relevant to gender equity intersects with governance and knowledge diffusion concerns, and includes examinations of women's low participation rates in all economic sectors (Wekwete 2014); structural impediments to women successfully completing higher education (Chipfuwa et al. 2014); concerns over government effectiveness, including enhancing access to development resources, such as access to water and sanitation (Wekwete and Manyeruke 2012; Jerie 2011); intimate partner violence faced by women (Wekwete 2014; Viriri et al. 2011); and how these issues contribute to women's disfranchised positions in Zimbabwe society.

Theme 1: gender inequality remains prevalent in all sectors of the economy, given both cultural norms and policy in Zimbabwe

Female and feminist/gender-conscious scholars in Zimbabwe generally agree that gender inequality is still a complex and prevalent social problem that exists in a variety of forms embedded in all sectors of the society. For instance, Dr. Naomi Netsayi Wekwete, a prolific scholar from the University of Zimbabwe, examines gender inequality as it relates to maternal and child wellbeing, adolescent reproductive health issues, HIV and AIDS intervention, migration, and economic development. In her article "Gender and Economic Empowerment in Africa: Evidence and Policy," Wekwete (2014) explains that the majority of women in Africa today are still excluded from the formal labor market, which is dominated by men. Deep rooted cultural norms that women are subordinate to men and that women should shoulder the responsibility for domestic work in the home reinforce male dominance. Therefore, some women in Zimbabwe have limited access to jobs in the formal labor market, to credit, and to land resources and other extension services. Under this context, women have to work in the informal sector or on small pieces of land to earn a meager wage, and often their agricultural labor is unpaid. According to Wekwete (2014), these inequalities are due to policies that reinforce gender inequities. Wekwete provides five critical recommendations to improve women's economic situation. First, labor regulations must grant women access to the formal labor market. Second, skills training must be offered to enable women to find jobs in the formal labor market. She then recommends the "setting up of micro-credit schemes, use of technology to access markets such as mobile phones to release women's time in caring and domestic work, [and] fostering of partnership by providing funding for women, cash transfers, and welfare fund" (2014, pp. i87–i88). Also, she suggests improving infrastructure, including water and electricity. Lastly, childcare must be subsidized in order to help release women from time-consuming housework and enable their work in formal labor markets.

In addition to gender inequities in the workforce, gender inequality is also sustained through cultural norms, including women's access to land and their empowerment in Zimbabwe. Some of the leading researchers in this area of study include Professor D. Z. Moyo's research team of Midlands State University (MSU), as well as authors who contributed to a special issue devoted to gender studies of MSU's peer-reviewed journal *The Dyke*. In it, Crescentia Madebwe and Victor Madebwe (2011) examine gender dynamics of Zimbabwe's fast-track land reform program, which limits women's ability to access, own, control, use, and manage land resulting from the political and ideological controversies surrounding its implementation. The authors provide evidence that gender inequality in the decision-making process, as well as the distribution of land and material resources, impedes Zimbabwe's achievement of their poverty reduction goals. Therefore, they determine that "a human rights-based framework based on international legal standards such as the right to livelihood and the right to equality may remove land from the private realm of the family into the sphere" (Madebwe and Madebwe 2011, p. 100). In that way, women in Zimbabwe will benefit from land reform based on gender equality.

Zimbabwe women and gender-conscious scholars suggest that the benefits of gender equality ideology in land reform policy should also extend to economic sectors and public management. Indicative of this approach, Steven Jerie's article "Gender and Solid Waste Management in the Informal Sector of Bulawayo, Zimbabwe" utilizes questionnaires to "assess the role of gender in solid waste management in the informal sector in the high-density suburbs of Bulawayo, the second largest city in Zimbabwe" (2011, p. 46). Jerie argues that "more solid waste is generated in enterprises operated by men than those operated by women" and that "there is generally poor management of waste in the home industries, however, there is a greater level of cleanliness in the enterprises run by women who engage in waste reduction practices such as waste picking and recycling" (2011, p. 46).

As such, Jerie suggests that the incorporation of a gender equality perspective into the solid waste management sector would result in improved effectiveness of the industry and would provide women in Bulawayo with more opportunities to work in the formal sector.

Theme 2: the marginalization of women impacts gender-based violence in Zimbabwe, which threatens women's social status and wellbeing

According to scholars, women's disfranchised position in Zimbabwean society, buttressed by contemporary cultural factors such as Christianity, is the antecedent to intimate partner violence. Scholar Naomi Netsayi Wekwete and others of the University of Zimbabwe published "The Association between Spousal Gender Based Violence and Women's Empowerment among Currently Married Women Aged 15–49 in Zimbabwe: Evidence from the 2010–2011 Zimbabwe Demographic and Health Survey" (2014). The findings suggest that "women who did not participate in decision-making at [the] household level were more likely to experience GBV [gender based violence] than those who do" (2014, p. 1413). Further, women who earned less than their spouses were more vulnerable to GBV than their peers, and women under the age of 20 were more likely to experience GBV than older age groups. Consequently, Wekwete pushes for the improvement of women's participation in decision-making as well as their economic and social emancipation.

The MSU research team also examined issues of domestic violence. For instance, Advice Viriri, Excellent Chireshe, and Tompson Makahamadze's "'Domestic Violence Act and the Apartheid of Gender': A Critical Analysis of the Perceptions of Christian Women in Masvingo Province of Zimbabwe" (2011) examines women's "silence" concerning GBV. They explain that this occurs not only due to historical sociocultural norms, but also from the local Christian churches' teachings that "disputes should be settled within the home and church" (2011, p. 34). In fact, "the Bible is perceived as protecting Church members against domestic violence by preventing such cases from occurring" (2011, p. 35). Yet, through interviews, they learned that since churches sometimes fail to deal with cases of domestic violence, it is necessary to advocate for the 2007 Domestic Violence Act so that women who live in predominantly Christian communities can be protected, as is the case for women living in Masvingo. Therefore, the authors recommend that local churches "educate their members on the relevance of the Domestic Violence Act as it seeks to complement their effort to eliminate domestic violence" (2011, p. 43).

Theme 3: impediments to women's access to higher education

In their article "Strengthening Research Capacity in Africa: Gender, Sexuality and Politics with a Strategic Focus on the Lives of Young Women," Dr. Naomi Netsayi Wekwete and Charity Manyeruke (2012), explain that female students' sexual and reproductive health and rights are made more vulnerable due to poor public management. Their participatory action research took place at the University of Zimbabwe from April 2010 to April 2011, when lack of water supplies to residences resulted in the closure of student dormitories on campus. Although this lack of access to accommodation on campus affected all students, it especially affected the sexual and reproductive health of female students. For instance, lack of accommodation and being forced to find expensive housing in town led many young women into prostitution and to subject themselves to older partners in exchange for a place to stay. Furthermore, young women found it difficult to find safe spaces on campus for leisure and studying, away from the penetrating male gaze, commentary, and harassment. They recommend that in order to protect female students' basic health rights, the university authorities must consider the impact on female students when making decisions about the campus and are further obligated to *act* in order to alleviate these challenges faced by female students.

Systematic exclusion

The work of these female and male gender-conscious African scholars can only increase transnational scholars' knowledge of Zimbabwe's social issues and potential solutions if we provide better links between them and intellectuals in other parts of the world. However, searches in popular U.S. health and social scientific databases such as Popline and Sociological Abstracts do not elicit results that include many of these authors' works. For instance, while examining Popline, a free demography and health database that is supported by the United States Agency for International Development (USAID), of the seven first authors whose work was discussed above, I found that four *do not have any* publications listed in Popline, whereas as many as 30 peer-reviewed articles are listed on the university website of one of these authors (though six is the modal number of publications of overlooked authors). See Table 2.1 for a summary.

Evidence of database omissions of the work of scholars in Zimbabwe underscores problems that UNESCO hopes to address. The UNESCO and International Social Science Council World Report in 2010 articulated twenty-first-century priorities and expounded the need to address knowledge divides between scholars in the Southern hemisphere and North Atlantic researchers (UNESCO and ISSC 2010). It explained that Africa may be hardest hit by the structural and epistemic disparities that produce the knowledge divides (Harrison 2016). As shown by the bibliographies of many Western researchers who write about health and population in Africa, these divides continue. In this chapter, I seek to offset the hegemony of North Atlantic social scientists writing about Africa without engaging with African scholars' work or African audiences.

Some correctives to these Western/North Atlantic-focused databases are South-South collaborations such as the Council for the Development of Social Science Research in Africa (CODESRIA) and journals such as *Feminist Africa* and *Jenda.* In addition to my searches of university websites and my use of Western databases, I utilize South-South resources as much as possible.

ICT activism in Zimbabwe

Guided by African feminist methodology, I seek not only to uncover the work of African scholars, but also to unveil the gems of understanding offered by activists who work in Zimbabwe. Women and their families who are working toward social change in Zimbabwe have created numerous organizations and pursued myriad strategies in order to challenge the status quo in their country and to provide for the well-being of their families. Below I discuss the ideas and strategies emanating from the scholarship and activism of women in Zimbabwe.

While examining literature, newspaper articles, and other sources, I (2014) found that women who are involved in nongovernmental

Table 2.1 Popline and Sociological Abstracts analysis of African women's scholarship cited in Western databases

Author	*Publications listed in Popline*	*Publications listed in Sociological Abstracts*	*Publications listed on university and other websites*	*Publications noted under review*	*Citations (Research Gate* and Google+)*	*Link*
Chipfuwa, Tirivanhu	1	1	9		23+	www.researchgate.net/profile/Tirivanhu_Chipfuwa2
Jaji, Rosemary	0	1	4	4	44+	www.uz.ac.zw/index.php/socio-staff/211-faculty-of-social-studies/departments/sociology/560-dr-rosemary-jaji
Jerie, Steven	0	1	7		33+	www.africaportal.org/search/site/jerie%2C%20s
Madebwe, Crescentia	0	2	9		16*	www.researchgate.net/profile/Crescentia_Madebwe
Moyo, Doreen Z.	0	0	9		44*	www.researchgate.net/profile/Doreen_Moyo/publications
Viriri, Advice	0	0	34		21+	http://ww4.msu.ac.zw/member/dr-a-viriri/
Wekwete, Naomi	2	2	20		39+	www.uz.ac.zw/index.php/staff-pop/209-faculty-of-social-studies/departments/population/675-dr-naomiwekwete
Mode:	0	1	9			
Mean:	0	1	13			

Source: Zerai, Perez and Wang (2016).

Table 2.2 A women-centered model for ICT, knowledge economy, and governance: variables in models and data sources

Variables (note: unit of analysis, women aged 15–49, with country variables matched to women-country-year file)	*Country Policy Analysis (CPIA)*	*IDHS: "IPUMS-DHS"*	*Worldwide Governance Indicators (WGI)*
ICT (concept and a variable)			1
Mobile-cellular subscriptions (per 100 people)			1
Percentage of women with mobile-cellular accounts			1
Household has cell phone		1	
Female respondent owns cell phone		1	
Household has smartphone with internet access		1	
Household has personal computer		1	
Respondent uses own cell phone for financial transactions		1	
Respondent used the internet ever and in past year		1	
Respondent-frequency of using internet in last month		1	
Knowledge diffusion (Asongu and Nwachukwu 2016)			
Formal education:			
Percentage of women completing primary education (educational quality)			1
Female respondent, highest educational level, secondary [or further]		1	
Woman's completed years of education		1	
Primary teacher-pupil ratio (educational quality)			1
School enrollment, primary (gross), gender parity index (GPI)			1
School enrollment, primary and secondary (gross), gender parity index (GPI)			1
Formal/informal education:			
Woman's reported literacy		1	
Woman's literacy, based on reading passage		1	
Woman's literacy bridging variable		1	
Respondent ever attended literacy program		1	
Respondent reads newspaper: bridging variable		1	
Respondent watches television: bridging variable		1	

Variables (note: unit of analysis, women aged 15–49, with country variables matched to women-country-year file)	*Country Policy Analysis (CPIA)*	*IDHS: "IPUMS-DHS"*	*Worldwide Governance Indicators (WGI)*
Respondent listens to radio: bridging variable		1	
General governance, including political, economic, and institutional governance			
Voice and accountability (political governance)			1
Women's involvement in government/ NGOs		1	
Percentage of women elected to office			1
Rule of law (institutional governance):			
Intimate partner violence (IPV) prosecutions/IPV is prosecuted			1
Government effectiveness (economic governance):			
Coverage of social safety net programs (percentage of population)	1		1
Equity in access to clean water, improved sources			
Improved water source (percentage of population with access)			1
Improved water source, rural (percentage of rural population with access)			1
Improved water source, urban (percentage of urban population with access)			1
Equity in access to improved sanitation			
Improved sanitation facilities (percentage of population with access)			1
Improved sanitation facilities, rural (percentage of rural population with access)			1
Improved sanitation facilities, urban (percentage of urban population with access)			1
CPIA policy and institutions for environmental sustainability rating (1=low to 6=high)	1		
CPIA public sector management and institutions cluster average (1=low to 6=high)	1		
Nondiscrimination clause mentions gender in the constitution (1=yes; 0=no)			1
Law mandates equal remuneration for females and males for work of equal value (1=yes; 0=no)			1

(*continued*)

Table 2.2 (cont.)

Variables (note: unit of analysis, women aged 15–49, with country variables matched to women-country-year file)	*Country Policy Analysis (CPIA)*	*IDHS: "IPUMS-DHS"*	*Worldwide Governance Indicators (WGI)*
Law mandates nondiscrimination based on gender in hiring (1=yes; 0=no)			1
CPIA policies for social inclusion/equity cluster average (1=low to 6=high)	1		
CPIA gender equality rating (1=low to 6=high)	1		
Control factors (individual level, with women aged 15–49 as unit of analysis):			
Urban-rural status		1	
IDHS sample identifier (country and year)		1	
Female respondent's age		1	
Female respondent's relationship to household head		1	
Sex of household head		1	
Relationship structure in household		1	
Woman's current marital or union status		1	
Household has telephone		1	
Attitude to wife beating: justified if woman refuses to have sex		1	
Barrier to woman's health care: lack of money for treatment		1	
Barrier to woman's health care: distance to facility		1	

Notes:
The *Country Policy and Institutional Assessment (CPIA)* of the World Bank, and adopted in a modified form by the African Development Bank (AfDB), "is a rating system designed to assess the performance of countries' policy and institutional frameworks in terms of their capacity to ensure the efficient utilization of scarce resources for achieving sustainable and inclusive growth. It is based on the scoring of 18 criterial covering different aspects of development such as: economic and public-sector management; structural policies; social inclusion and equity" (African Development Banks 2015, p. 3). CPIA data downloads are derived from the World Bank website (World Bank Group, CPIA database [accessed at: www.worldbank.org/ida]).

IDHS is an IPUMS (the Integrated Public Use Microdata Series Group that provides census and survey data at the University of Minnesota Population Center)-DHS partnership that provides aggregate DHS data.

WGI from the World Bank (Kaufmann et al. 2010) capture six key dimensions of governance (voice and accountability, political stability and lack of violence, government effectiveness, regulatory quality, rule of law, and control of corruption) for 200 countries between 1996 and the present day.

organizations (NGOs) are a leading resource for carrying out the activism of women. Women have started their own organizations and have also tapped into global nongovernmental networks, such as Amnesty International, Physicians for Human Rights, and other forms of support for their efforts.

ICT and Women of Zimbabwe Arise (WOZA)

An example that excites the imagination is the organization Women of Zimbabwe Arise.

In the face of the realities of living in a repressive regime, two courageous women, Sheba Dube and Jenni Williams, came together in 2002 to begin discussions and make plans for a movement of mothers, grandmothers, sisters, and aunts dedicated to a better future for the people of Zimbabwe. Together, they formed WOZA (Women of Zimbabwe Arise, later altered to Women and Men of Zimbabwe Arise).

In addition to their nonviolent protest, which included "road writing," where WOZA activists "write our message of love on the street" and "a very high-risk model of going and sitting at a government complex and demanding that they declare a national famine," WOZA spearheaded the development of a People's Charter. It is there where activists and others clarify "what we think of health, what we think we need in terms of educating our children, righting the wrongs in the history of the Zimbabwe transitional justice issues" (from an interview with human rights defender Jenni Williams; Amnesty International 2008).

The People's Charter has mass appeal because its crafting took into account ordinary citizens' demands. The WOZA website explains their process:

> Since January 2006, WOZA has carried out consultations on social justice across the country. In 284 meetings, almost 10,000 rural and urban people told us what they want in a new Zimbabwe. We wrote down what they said and the result is the People's Charter.

The government of Zimbabwe did not respond favorably to the charter. "And for launching that people's charter, we were battered and bruised and put in jail" (from an interview with human rights defender Jenni Williams; Amnesty International 2008). ICT were important to WOZA's efforts because the organization was able to gain international notoriety resulting from the availability of interviews with Jenni Williams on Amnesty International's website. In addition, ICT support the continued dissemination of the People's Charter through WOZA's website.

Solutions: a recap of African feminist lessons

Many solutions are offered by the thoughts and actions of women's organizational, strategic, and academic efforts that support the principles of an African feminist approach. They advocate for widespread education, emphasizing women's and girls' access to education (Wekwete et al. 2012; WUA/UNICEF Partnership 2015), and access to reproductive health information and resources (Wekwete et al. 2012, Chipfuwa et al. 2014). Also, they call for a robust civil society and the necessary freedom for social movements to flourish (Amnesty International 2008). Clearly there is a need for a functioning and vibrant infrastructure for ICT, women's equity, and good governance. Zimbabwean scholars provide evidence of the need for economic opportunities for women (Madebwe and Madebwe 2011; Wekwete 2014), as well as for dialogue about and the redressing of gender inequity (Jaji 2009) and intimate partner violence (Viriri et al. 2011; Wekwete et al. 2014). Though gender inequality is pervasive in Zimbabwe and affects women intersectionally, as experiences differ by age, ethnicity, class, and social roles, Zimbabwe's future is bright, especially if the wisdoms offered by the women of Zimbabwe inspire praxis and social change.

The need for a Zimbabwean scholars and activists database

By creating a database, I hope to contribute to intellectual diversity in multiple fields.[5] As argued above, the words, theoretical perspectives, research, and political activity of women in African contexts have been rendered inaccessible to Western researchers. Their important potential contributions remain isolated due to the difficulties and resources needed in order to publish their work in Western and top-tier international journals. It is our task to break down these barriers and bring African women's efforts—both academic and activist—to the fore. After examining the online curricula vitae of female university scholars in Zimbabwe (as of November 2017), I gathered a database listing all of their publications and papers. Further, I perused online news sources and websites to garner women's understandings of gender equity and good governance in Zimbabwe. Next, I describe the dynamic database I have built of these factors from existing data sources by merging DHS data with World Bank and Country Policy Institutional Analysis indicators to address women scholars' and activists' perspectives on these issues in Zimbabwe. See Table 2.2 for a list of variables and data sources for subsequent analyses.

World Bank definitions of governance

In *The Worldwide Governance Indicators: Methodology and Analytical Issues*, Daniel Kaufmann of the Brookings Institution, and his coauthors Aart Kraay and Massimo Mastruzzi (2010) of the World Bank describe the WGI project, which provides "governance indicators for over 200 countries and territories over the period 1996–2016, for six dimensions of governance": A. political governance, to include: 1. political stability and absence of violence/terrorism; and 2. voice and accountability: B. economic governance, to include: 3. regulatory quality, and 4. government effectiveness; and C. institutional governance, to include: 5. control of corruption; and 6. rule of law. In my intervention, I ask whether adding variables that highlight women's access improves the predictive power of this framework for women in Zimbabwe.

For each of these dimensions, plenty of data are available that capture women's status and experiences that could be added. Below and throughout this book, I draw from the literature of scholarly and activist women in Africa and the African Diaspora in order to make recommendations about the variables to consider.

New definitions of the diffusion of knowledge in Zimbabwe

In reexamining issues of knowledge diffusion and good governance that have been studied in the past from androcentric perspectives, the work of female scholars in Zimbabwe sheds light on a number of factors that may expand our understandings of how to measure knowledge diffusion, and gender equity, as part and parcel of good governance. In terms of knowledge diffusion, in Asongu and Nwachukwu's (2016) analysis, knowledge diffusion is examined through three factors: (a) the percentage of mobile-cellular telephone subscriptions per 100 in the population; (b) educational quality as measured by the primary teacher–pupil ratio; and (c) innovation, as measured by patents and trade-mark applications (see Figure 2.1). On the basis of the work of female scholars, it is important to examine knowledge diffusion *among women*. So, an expanded definition of educational quality that includes the percentage of women completing primary and secondary or further education is in order. Another possible measure would include the ratio of girls to boys in school (i.e. the gender parity index) (see Figure 2.2).

New definitions of good governance in Zimbabwe

In Asongu and Nwachukwu's (2016) analysis, four dependent variables are examined. Each taps into an important aspect of good governance.

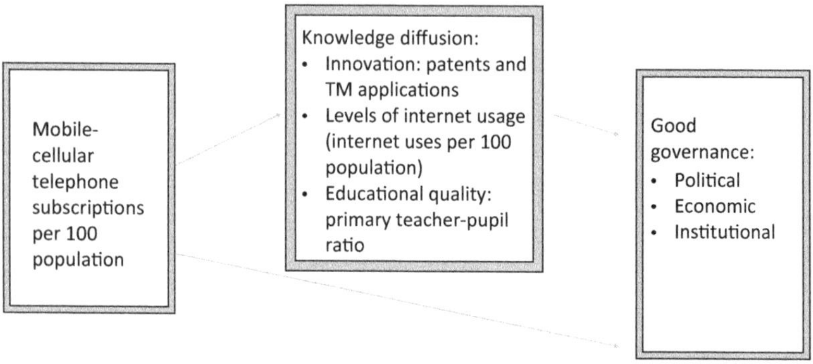

Figure 2.1 Conceptualizing the ICT and knowledge diffusion hypothesis in Asongu and Nwachukwu's model (2016)

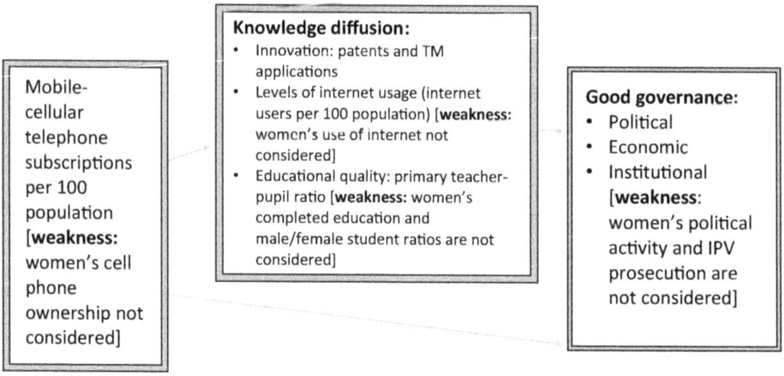

Figure 2.2 Conceptualizing weaknesses in the ICT and knowledge diffusion hypothesis

These include political governance (a combination, through the creation of a Principal Components Analysis [PCA] factor, of political stability, lack of violence, and voice and accountability), economic governance (a PCA factor comprised of government effectiveness and regulation quality), institutional governance (a combination [via PCA] of the rule of law and corruption control), and, finally, a fourth dependent variable that combines political, economic, and institutional governance into a PCA factor that Asongu and Nwachukwu (2016) define as general governance. All four factors are crucial to the wellbeing of women and their families. And the literature demonstrates the importance of ICT to political governance, and in particular the voice and accountability dimension of this factor. For example, as noted above, ICT supported the work of developing and disseminating the People's Charter in Zimbabwe. Currently, it is available on the WOZA website to download in English and multiple national and local languages. Further, in posting an interview of WOZA activist Jenni Williams, Amnesty International drew attention to her 33 arrests resulting from her efforts as a human rights defender.

The necessity of good economic governance, including the government effectiveness dimension of this factor, is also highlighted in women scholars' work. Clean water, safe sanitation, and access to nearby health facilities, including good roads to travel to them, are all aspects of good economic governance. And clearly the rule of law and corruption control are relevant to women as aspects of institutional governance.

While political, economic, and institutional governance contribute to the wellbeing of women, on the basis of the literature offered by women academics in Zimbabwe, to the three factors making up general governance, I would offer gender equity as an additional factor.

Measures of gender equity that are consistent with the intellectual work of female scholars include average levels of women's completed education, women's employment opportunities, women's involvement in government, including the percentage of elected officials who are women, and whether domestic violence is prosecuted in the country.

Summary: building a new model for analyzing World Bank and DHS data to determine impact of cell phones, and diffusion of knowledge on good governance and gender equity

In Asongu and Nwachukwu's (2016) work, mobile-cell telephone subscriptions per 100 of the population is the main independent variable under examination, along with the mechanism of knowledge diffusion through educational quality, innovation, and internet coverage. In the spirit of the literature offered by women scholars in Zimbabwe, I make one additional suggestion that we expand the mobile-cellular telephone subscriptions variable to encompass the percentage of women with mobile-cellular accounts.

In summary, in the new model (Figure 2.3), I propose to expand the analysis offered by Asongu and Nwachukwu (2016) to include the following:

1. Instead of three governance indicators (political, economic, and institutional governance) as dependent variables, I propose that gender equity is added as a dimension of good governance.

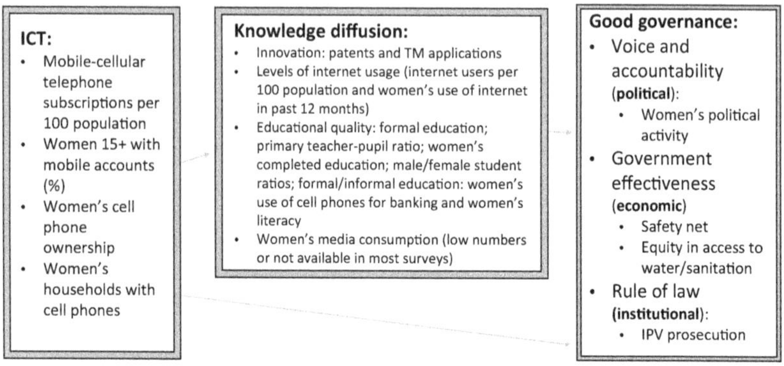

Figure 2.3 Zerai's women-centered model for mobile use, knowledge diffusion, and governance

2. The main independent variables will be mobile-cellular telephone subscriptions per 100 of the population and the percentage of women with mobile-cellular accounts.
3. Two of the three knowledge diffusion variables (innovation and internet coverage) will remain the same as in Asongu and Nwachukwu's (2016) conceptualization. The third, educational quality, will be expanded to include not only teacher-per-pupil ratios, but also the percentage of women who have completed primary education, the percentage of women who have completed secondary education, and ratio of girls to boys in primary education and in secondary education. In my subsequent analyses (see Chapter 5), these variables will be examined separately and as a PCA factor.
4. In conceptualizing political governance, whether domestic violence is prosecuted will be examined, along with World Bank indicators of perceptions of domestic violence that are already in Asongu and Nwachukwu's (2016) measure.

Coda

This conceptualization is fascinating theoretically, but what does it look like in practice? Figure 2.4 provides a depiction of a model that operationalizes the concepts named above for the Zimbabwe context. In previous work (Zerai 2014; Henderson, Zerai, and Morrow 2017; Zerai et al. 2018), I have shown that ethnic cleavages are extremely

Table 2.3 Number and percentage of women aged 15–49 who own cell phones, by region in Zimbabwe (2015)

Region	*Number*	*Percentage*
Manicaland	847	66.9
Mashonaland Central	473	53.6
Mashonaland East	613	64.4
Mashonaland West	714	61.6
Matabeleland North	286	61.5
Matabeleland South	313	74.5
Midlands	874	69.3
Masvingo	815	68.7
Harare	1,483	83.2
Bulawayo	500	86.7
Total (out of 9,955)	6,918	69.5

Source: ZDHS (2015).

Table 2.4 Descriptive statistics, unit of analysis: women aged 15–49, Zimbabwe (2015)

Variable	*Number*	*Mean or proportion*	*Standard deviation*
ICT			
Household has landline telephone	9,955	0.36	1.471
Household has cell phone	NA	NA	NA
Female respondent owns cell phone*	9,955	0.69	0.46
Respondent uses own cell phone for financial transactions*	6,918	0.54	0.49
Knowledge diffusion			
Literacy—reads passage easily	9,955	0.87	0.33
Literacy—reads passage with difficulty	9,955	0.94	0.22
Female respondent, highest educational level, secondary	9,955	0.08	0.28
Female respondent, primary education	9,955	0.78	0.41
Female respondent, incomplete primary or no formal education	9,955	0.12	0.33
Government effectiveness/development resources availability			
Drinking water on premises	9,955	0.37	0.48
Unshared improved toilets	9,955	0.38	0.48
Barrier to woman's health care: lack money for treatment is a big problem	9,955	0.42	0.49
Barrier to woman's health care: distance to facility is a big problem	9,955	0.33	0.47

(continued)

Table 2.4 (cont.)

Variable	*Number*	*Mean or proportion*	*Standard deviation*
Control variables	9,955		
Urban residence	9,955	0.38	0.48
Age of female respondent (15–49)	9,955	28.53	9.29
Resident is a visitor (versus usual resident)	9,955	1.05	0.21
Woman is currently married	9,955	0.58	0.49
Respondent agrees wife beating is not justified	9,955	0.83	0.36
Region/zone			
Shona-dominant region	9,955	0.48	0.49
Ndebele-dominant region	9,955	0.14	0.35

Source: Zerai's analysis of TDHS (2015).

* Data collected in last DHS only (Zimbabwe 2015).

Table 2.5 Determinants of ownership of cell phones: women aged 15–49, Zimbabwe (2015)

Variables in the equation	*B*	*S.E.*	*Wald*	*df*	*Sig.*	*Exp (B)*	*95% C.I. for Exp(B)*	
							Lower	*Upper*
Shona-dominant region	–0.307	0.051	35.747	1	0.000	0.735	0.665	0.813
Ndebele-dominant region	0.152	0.075	4.064	1	0.044	1.164	1.004	1.350
Woman is currently married	0.613	0.047	168.339	1	0.000	1.846	1.683	2.025
Literacy—reads passage easily	0.909	0.064	204.445	1	0.000	2.481	2.191	2.810
Urban residence	1.355	0.055	598.101	1	0.000	3.876	3.477	4.320
Constant	–0.608	0.072	70.637	1	0.000	0.545		

Source: Zerai's analysis of ZDHS (2015).

Note: B = coefficient; S.E. = standard error; Wald = Wald statistic; df = degrees of freedom; Sig. = significance level; Exp(B) = odds ratio; C.I. = confidence interval.

Table 2.6 Summary of impact of a woman's ownership of a cell phone on respondent's literacy, access to drinking water, unshared flush toilets, cost barriers, distance barriers, views on IPV: women aged 15–49, Zimbabwe (2015)

Variables in the equation	*Drinking water on premises*		*Unshared flush toilets at home*		*Cost is a barrier to woman's health care*		*Distance is a barrier to woman's health care*		*Disagreement that IPV is justified if wife refused to have sex*	
	Exp(B)	*Sig.*	*Exp(B)*	*Sig.*	*Exp(B)*	*Sig.*	*Exp(B)*	*Sig.*	*Exp(B)*	*Sig.*
Shona-dominant region	.523	***	1.076		.990		.962		.959	
Ndebele-dominant region	1.097		1.185	*	.875	*	1.072		1.823	***
Literacy	1.324	***	1.703	***	.577	***	.656	***	1.591	***
Woman is currently married	.853	*	.631	***	1.064		1.289	***	1.058	
Urban residence	9.948	***	1.802	***	.555	***	.139	***	1.818	***
Female respondent owns cell phone	1.384	***	1.395	***	.828	***	.755	***	1.398	***
Constant	.203	***	.297	***	1.702	***	1.320	***	2.108	***

Source: Zerai's analysis of ZDHS (2015). This table simplifies findings from the analysis of seven dependent variables. Detailed results are available upon request.

Notes:

B = coefficient; Sig. = significance level; Exp(B) = odds ratio.

Significance levels are grouped into three categories: *$p<0.05$; **$p<0.01$; ***$p<0.001$.

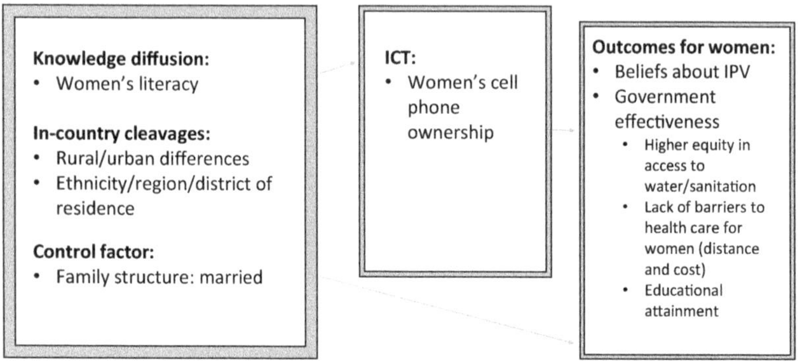

Figure 2.4 Ownership of cell phones and individual outcomes for women aged 15–49, Zimbabwe

important in Zimbabwe. They contribute to perspectives on experiences of intimate partner violence (Henderson, Zerai, and Morrow 2017), access to water and sanitation (Zerai 2014; Zerai et al. 2018), to access to health care (Zerai 2014), and to educational attainment (Zerai 2014). In Chapter 1, I demonstrated that cell phones in women's household are related to knowledge diffusion and the women's outcomes delineated in Figure 2.4. So, this provides some support for the veracity of my theory that women's cell phone ownership and knowledge diffusion among women can lead to better governance for all people.

Taking a brief look at data from the Zimbabwe DHS, Tables 2.3 through 2.6 provide details concerning women's ownership of cell phones by province, descriptive statistics for relevant variables, as well as the logistic regression results of analyses of determinants of both women's cell phone ownership, and the impact of cell phone ownership on a number of women's outcome variables. Given the high number of cell phones owned by women in Zimbabwe, the range by region shows high percentages of ownership across the board, ranging from 54% to 87% of women aged 15-49 in 2015 (Table 2.3). Descriptive statistics (Table 2.4) provide a comprehensive picture of factors noted in the models proposed in this chapter. I grouped regions according to the dominant languages in them, and found that 48% of women live in Shona dominated regions, whereas 15% live in Ndebele dominated ones. These regional differences have a remarkable affect on women's cell phone ownership (Table 2.5). Women in Shona dominated provinces are less likely to own cell phone relative to women in other provinces. And women in Ndebele-dominant provinces are more likely to own cell phones. And urban residence has an expected positive effect; women living in urban areas are 3.88 times as likely to own cell phones relative to women in rural areas of Zimbabwe (Table 2.5).

Table 2.6 summarizes the impact of cell phone ownership and a number of other factors on women's outcomes, as noted in this chapter. It does not include cell phone ownership's positive association with literacy and higher levels of education, though these results were found. Women's cell phone ownership in Zimbabwe is also associated with development resources such as drinking water on premises, and improved sanitation. And it is negatively related to women experiencing cost and distance as barriers to obtaining their own health care. Finally, women's cell phone ownership is related to women's disagreement that IPV is justified if a wife refused to have sex (see Table 2.6). Next, in order to test these ideas beyond the Zimbabwe context, I examine literatures and data from two different contexts, Tanzania and Malawi. Also, I operationalize models to test my general theory.

In the Introduction and Chapters 1 and 2, I critically examined factors that the World Bank uses to describe good governance and called for an intervention by African women scholars. I applied an African feminist lens to the study conducted by Asongu and Nwachukwu (2016) and found: (1) their model analyzing the impact of ICT on knowledge economies and good governance lacks factors that consider issues of gender equity in the definition of good governance; and (2) ICT access and knowledge diffusion must be measured in ways that consider women's access. In the following chapters, I continue to build on my initial work to create a more inclusive definition of good governance, and I conduct multivariate analyses of the impact of ICT on good governance in Tanzania and Malawi.

Notes

1 Parts of Chapter 2 were previously published in *Qualitative Inquiry* (Zerai et al. 2016). Appropriate permissions have been obtained.

2 Given that I (Zerai 2014) have already published a review of their work, herein I draw attention to the scholarship of lesser-known academics in North Atlantic contexts.

3 Not all universities contained faculty listings by department. An additional limitation is that even for faculty listed, many did not have personal websites, c.v.'s, or other evidence of electronic footprint available.

4 Given that the female database is dynamic and large, I provide a stable URL for it in lieu of providing a static listing: www.researchgate.net/project/African-Women-Scholars-Database?_updateId=5b73a8c6cfe4a7f7ca5a5f70&_pua=true.

5 The literature that this work addresses includes the sociology of sex and gender (Guillaumin 1995; Risman 2004); race, gender, and class (intersectionality) (Collins 1990; Uttal and Cuádraz 1999; Zerai and Banks 2002); the sociology of health and illness as a branch of medical sociology (Charmaz and Paterniti 1999; Conrad 2003); health activism related to both social movements and the sociology of health and illness (Zerai and Banks 2002; DeLaet and DeLaet 2012; Zerai 2014); methodology (Zerai and Banks 2002; Zerai 2014); the sociology of knowledge (Rabaka 2010); African (Gaidzanwa 1992, 1993, 1996, 1997, 2006, 2013) and African-American sociology (Du Bois 1897, 1899, 1903; Cooper 1998; Collins 1990; Rabaka 2010); and the transdisciplinary development effectiveness literature (Brautigam 2009).

3 ICT, women's status, and governance in Tanzania, 2010 and 2015–2016

The United Republic of Tanzania has articulated and implemented policies to reduce poverty and improve the standard of living for families for many decades. Their work has also included a focus on women's empowerment for some time. The rise in cell phone ownership has accompanied improvements in access to a number of development resources that may reflect strengthened government effectiveness. In the country overall, cell phones in households have increased from 46% in 2010 to 78% in 2015–2016 (Tanzania Demographic and Health Survey 2011, 2016). A deep dive into regional and rural-urban differences makes for an engaging discussion of the varied ways in which women's status, diffusion of knowledge, and quality governance are represented throughout the country.

In this chapter, I examine organically determined definitions of these issues that emanate from women scholars and activists in Tanzania. I then build and analyze a dataset that centers these locally relevant understandings of gender equity, knowledge economies, and the mobile ecosystem in order to analyze the impact of women's ownership of and access to a cell phone in their households on their knowledge and governance indicators that takes differences between households and among varied regions within Tanzania into account. The asymmetry in access to cell phones, knowledge networks, and development resources for women in Tanzania features prominently in this chapter. In the 2004–2005 Tanzania Demographic and Health Survey (TDHS), 9% of women's households owned landline or cell phones. And in the 2011 TDHS, the question about cell phones was asked separately. So, we can learn that almost half of households had cell phones. But as I argued in the Introduction, knowing that a cell phone exists in the household gives us no sense of *women's* access to cell phones. If we want to think about the impact of cell phones on women's empowerment and their voice in achieving good governance, or the ways in which ICT enhance government effectiveness for women, it is important to obtain laser-focused data on *individual* women's ownership of cell phones. Such data that include other descriptive factors and community-level and country-level indicators of women's status are extremely rare. Fortunately, the DHS provide for access to nationally representative samples of women aged 15–49. So, in the analyses in Chapters 2–5, I focus on the 2015 and 2015–2016 datasets from the Zimbabwe, Tanzania, and Malawi DHS that include questions that move beyond household access to cell phones and that embrace women's individual ownership. In Tanzania, for example, while cell phones are estimated to be found in 78% of households in 2015–2016, 52.2% of women aged 15–49 own cell phones.

As I have shown elsewhere (Zerai, Morrow, and Cuthbertson 2018), ethnic cleavages are not as important in Tanzania compared to Zimbabwe. However, residence in higher-resourced regions of the country is extremely relevant. This especially includes living in Zanzibar, Kilimanjaro, and Dar es Salaam. I also examined the differences between living on Pemba Island, in Arusha, and in other town/urban areas, and living in rural areas that dominate the rest of the country (covering 70% of Tanzania in 2012). However, these differences were not statistically significant for most indicators.

In the following pages, I examine the literature emanating from women of African descent in Tanzania and summarize their concerns regarding access to ICT, women's status, knowledge diffusion, and aspects of good governance that inform my analysis of the TDHS.

Literature by women scholars and concerning women's status and access to ICT in Tanzania

There exists prolific scholarly literature concerning cell phones in Tanzania. My most recent search turned up 523 results on one comprehensive search engine. When I narrowed these results down to articles that also mention gender in the title or as a keyword, the number fell to 15.

In addition to these 15, I did a search of websites from female scholars from Tanzania.

Themes in the literature include: use of mobile payment and banking (Lwoga and Lwoga 2017; Chilimo 2010), barriers to mobile battery recycling (Tedre et al. 2013), cell phones and diffusion of knowledge (Isaya et al. 2018), cell phones and health (Lund et al. 2012; L'Engle 2013; Linde et al. 2017), gender and foundations for STEM education (Kabote et al. 2014), and subject matter beyond the scope of this manuscript such as transactional sex and cell phones (Stark 2013), and gender stereotypes in media (Kahamba and Sife 2017).

A prominent area in the literature is that of the use of cell phones to promote health. This is a powerful theme because knowledge diffusion is the mechanism by which cell phones promote health. For example, Lund et al. (2012) report their study of an "mHealth project in Zanzibar called Wired Mothers [which] sent pregnant women text messages with appointment reminders and health education information, and it also gave enrollees phone vouchers to call health care providers directly with questions." They found that "the project significantly increased facility-based births among urban women but had no significant effect on rural women" (Lund et al. 2012, p. 11). And Linde et al.'s (2017) randomized trial of HPV-positive women in Tanzania showed that text messages increased the number of women who obtained follow-up cervical screening.

While the Tanzania DHS data sets do not have information on the use of phones for medical reminders, it does have data on various barriers to health care. So, I examine factors related to whether or not women experience cost or distance barriers to obtaining their own health. Mobile banking is also prominent in the literature, and I am able to examine descriptive information from the DHS in that regard. The relationship between a woman's ownership of a cell phone and knowledge diffusion is examined below.

A new conceptual framework

In Figure 3.1 I depict a new conceptual framework on the basis of the literature cited above. Women's literacy is my proxy for knowledge diffusion. Its impact is not entirely clear because cell phone ownership is shown both to contribute to literacy (Sanya and Odero 2017) as well as to be more effective when the user already reads well (Crawford et al. 2014).

Therefore, I use literacy in my models both to examine predictors of cell phone ownership, and women's access to health care and development infrastructure. In addition to literacy, I hypothesized that in-country cleavages will be evident (as I have shown in previous work [Zerai 2014; Zerai, Morrow, and Cuthbertson 2018]) in locales where governments are not as effective at delivering equitable services. The differences I examined are rural/urban area, and region, as described above. Marriage is added as a control factor.

First, I examined the impact of these factors on women's cell phone ownership. Then I examined the impact of these factors, in addition to

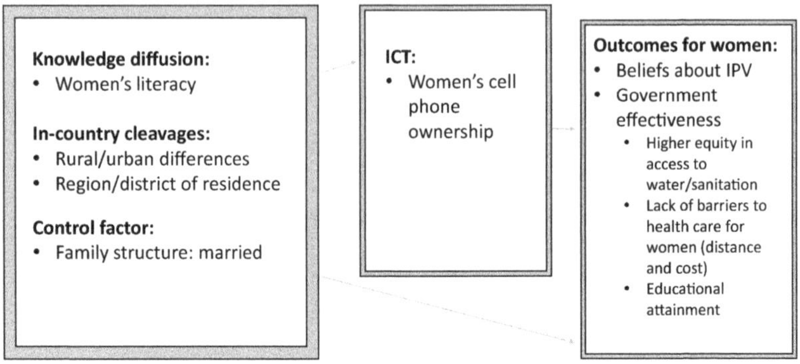

Figure 3.1 Ownership of cell phones and individual outcomes for women aged 15–49, Tanzania

women's cell phone ownership, on various outcomes relevant to women (as shown in Figure 3.1). These include her views on "wife beating" (i.e. intimate partner violence), and on outcomes in individual women's lives that signal general government effectiveness, including higher equity in access to water/sanitation, lack of barriers to health care for the respondent (distance and cost), and her educational attainment.

Describing the data

Women's ownership of cell phones in various regions of Tanzania

Table 3.1 provides a breakdown of women's ownership of cell phones by region. The areas that stand out are Arusha, Mjini Magharibi (in Zanzibar), Kilimanjaro, and Dar es Salaam. Over 70% of women own cell phones in each of these areas, distinguishing them from the rest of the country in which women's cell phone ownership ranges from 26% to 64% (averaging 52% among all regions). This will be instructive when I examine differential access to cell phones in logistic regression analysis. In the meantime, chi-square analysis results indicate that there are highly statistically significant differences between regions of Tanzania in women's ownership of cell phones (results not shown).

Women's educational attainment, literacy, and access to health, water, and sanitation

Women's ICT ownership and access, literacy, educational attainment, access to water, sanitation, health, residence, age, marital status, and

Table 3.1 Number and percentage of women aged 15–49 who own cell phones, by region in Tanzania (2015–2016)

Region	*Number*	*Percentage*
Mainland:		
Dodoma	222	38.8%
Arusha	360	71%
Kilimanjaro	270	74.8%
Tanga	450	63.7%
Morogoro	326	51.3%
Pwani	177	62.1%
Dar Es Salaam	1,306	85%
Lindi	145	50.3%
Mtwara	217	52.7%
Ruvuma	166	46.1%
Iringa	143	58.4%
Mbeya	428	51.7%
Singida	167	45.1%
Tabora	272	36.9%
Rukwa	75	26%
Kigoma	198	36.5%
Shinyanga	210	41.7%
Kagera	256	41.8%
Mwanza	359	41.8%
Mara	233	44.6%
Manyara	192	48.7%
Njombe	125	61.6%
Katavi	45	34.6%
Simiyu	157	32.8%
Geita	158	32.6%
Unguja (Zanzibar Island):		
Kaskazini Unguja	29	51.8%
Kusini Unguja	22	62.9%
Mjini Magharibi	155	77.1%
Pemba Island:		
Kaskazini Pemba	30	53.6%
Kusini Pemba	27	49.1%
Total (out of 13,264)	6,920	52.2%

Source: TDHS (2015–2016).

Table 3.2 Descriptive statistics, unit of analysis: women aged 15–49, Tanzania (2015–2016)

Variable	*Number*	*Mean or proportion*	*Standard Deviation*
ICT			
Household has landline telephone	12,561	0.010	0.083
Household has cell phone	12,561	0.850	0.357
Female respondent owns cell phone*	13,266	0.520	0.500
Respondent uses own cell phone for financial transactions*	6,921	0.700	0.456
Knowledge diffusion			
Literacy—reads passage easily	13,266	0.724	0.447
Literacy—reads passage with difficulty	13,266	0.044	0.205
Female respondent, highest educational level, secondary or further	13,266	0.138	0.344
Female respondent, incomplete primary or no formal education	13,266	0.264	0.441
Government effectiveness/development resources availability			
Drinking water on premises	13,266	0.295	0.456
Unshared improved toilets	13,266	0.228	0.420
Barrier to woman's health care: lack of money for treatment a big problem	13,266	0.423	0.494
Barrier to woman's health care: distance to facility a big problem	13,266	0.495	0.500
Control variables			
Urban residence	13,266	0.363	0.481
Age of female respondent (15–49)	13,266	28.690	9.652
Resident is a visitor (versus usual resident)	13,266	0.053	0.224
Woman is currently married	13,266	0.449	0.497
Respondent disagrees wife beating is justified if woman refuses to have sex	13,266	0.661	0.473
Region/zone			
Dar es Salaam	13,266	0.116	0.320
Kilimanjaro	13,266	0.027	0.163
Arusha	13,266	0.038	0.192
Zanzibar	13,266	0.022	0.147
Pemba	13,266	0.008	0.091

Source: Zerai's analysis of TDHS (2015–2016).

* Data collected in last DHS only (Malawi 2016, Tanzania 2015, and Zimbabwe 2015).

view of IPV are described in Table 3.2. Women's ICT ownership and access have been described above. And we will examine women's use of cell phones for financial transactions in Chapter 5. In relation to knowledge diffusion, the variables that are most important to scholars are literacy and education, especially including and beyond the secondary level. Almost three-quarters of respondents read a passage easily in English or Swahili (72%). However, a small proportion of women are still completing their secondary education and beyond (14%).

I examine government effectiveness at the individual level to see what impact the state's investments in development resources has on the lives of women. Less than one-third of women had drinking water on their premises in 2015–2016 in Tanzania. And 23% have access to unshared improved toilets at their homes. And both money and distance are barriers to women obtaining health care to 42% and 50% of women, respectively. A number of control variables also highlight women's empowerment. Less than half, 45% of women are married, and the majority of all women agree that wife beating is not justified if a woman refuses to have sex (66%). The average age of women in the dataset was 29.

In examining differential access to ICT (i.e. cell phones) and other analyses below, I distinguish Dar es Salaam as the only region that is 100% urban, from Kilimanjaro, Zanzibar, and the rest of the country (which is 70% rural, as noted above). As shown in Table 3.2, out of 13,266 female respondents, 12% are from Dar es Salaam, 3% from Kilimanjaro, 2% are from Zanzibar, 1% from Pemba Island, and 4% from Arusha. Whereas 62% are from rural areas on the mainland, and the rest are from urban mainland areas.[1]

To ease readability of chapters, details of reports of statistical analyses are streamlined in the text below. The unit of analysis for all logistic regression analyses in chapters 3 and 4 are women aged 15–49 in the Demographic and Health Survey samples.

Differential access to ICT (i.e. cell phones)

Table 3.3 provides results of a logistic regression analysis of women's ownership of cell phones in Tanzania from the 2015–2016 TDHS. In Table 3.3, I show that women's access to ICT (i.e. cell phones) is dependent upon living in more highly resourced regions of Tanzania. Given that only 8% of women used the internet in the past 12 months, I focus on women's cell phone ownership. Logistic regression analysis shows that women from Zanzibar are 2.2 times as likely to own a cell phone, women living in Kilimanjaro are 2.9 times as likely to own a cell

Table 3.3 Determinants of ownership of cell phones: women aged 15–49, Tanzania (2015–2016)

Variables in the equation	*B*	*S.E.*	*Wald*	*df*	*Sig.*	*Exp(B)*	*95% C.I. for Exp(B)*	
							Lower	*Upper*
Zanzibar	0.788	0.135	33.9	1	0.000	2.199	1.687	2.868
Kilimanjaro	1.048	0.127	67.8	1	0.000	2.853	2.223	3.661
Dar es Salaam	0.952	0.082	134	1	0.000	2.59	2.204	3.044
Literacy—reads passage easily	1.012	0.044	528	1	0.000	2.751	2.524	2.999
Woman is currently married	0.435	0.039	124	1	0.000	1.545	1.431	1.667
Urban residence	1.132	0.046	616	1	0.000	3.101	2.836	3.391
Constant	−1.366	0.045	922	1	0.000	0.255		

Source: Zerai's analysis of TDHS (2015–2016).

Note: B = coefficient; S.E. = standard error; Wald = Wald statistic; df = degrees of freedom; Sig. = significance level; Exp(B) = odds ratio; C.I. = confidence interval.

phone, and those living in Dar es Salaam are 2.6 times as likely to own a cell phone, relative to women living in other regions of Tanzania. And those who are literate are 2.8 times as likely to own a cell phone relative to women who cannot read a passage easily. Women who were married at the time of the interview, were 1.55 times as likely to own a cell phone relative to women who were not married. And women living in urban areas were 3 times as likely to own a cell phone, relative to women living in rural areas. All results are highly statistically significant ($p<.001$). And confidence intervals are shown in all tables.

To summarize, women from Zanzibar, Kilimanjaro, and Dar es Salaam are more likely to own cell phones, as are women who are literate, married, and living in urban areas. All results are highly statistically significant ($p<.0001$).[2]

Knowledge diffusion

The knowledge diffusion variables I examine are literacy and secondary or a higher level of education. Literacy is measured by women who have completed secondary or further education or who can read a passage easily. Variables associated with literacy include women living in Zanzibar, Kilimanjaro, and Dar es Salaam or other urban areas of Tanzania (Table 3.4). Women who own cell phones are 2.736 times as likely to read a passage easily. And women who are married are only

Table 3.4 Determinants of literacy: women aged 15–49, Tanzania (2015–2016)

Variables in the equation	*B*	*S.E.*	*Wald*	*df*	*Sig.*	*Exp(B)*	*95% C.I. for Exp(B)*	
							Lower	*Upper*
Zanzibar	0.773	0.175	19.5	1	0.000	2.166	1.537	3.053
Kilimanjaro	1.07	0.178	36.2	1	0.000	2.916	2.058	4.132
Dar es Salaam	0.303	0.094	10.4	1	0.001	1.354	1.126	1.628
Woman is currently married	–0.396	0.041	91.8	1	0.000	0.673	0.621	0.73
Urban residence	0.657	0.055	145	1	0.000	1.929	1.733	2.147
Female respondent owns cell phone	1.006	0.044	520	1	0.000	2.736	2.509	2.983
Constant	0.433	0.033	168	1	0.000	1.542		

Source: Zerai's analysis of TDHS (2015–2016).

Note: B = coefficient; S.E. = standard error; Wald = Wald statistic; df = degrees of freedom; Sig. = significance level; Exp(B) = odds ratio; C.I. = confidence interval.

.673 times as likely to be literate.[3] So marriage has a negative association with literacy.

As with literacy, Table 3.5 shows that secondary or further education is associated with living in Zanzibar, Kilimanjaro, Dar es Salaam, and other urban areas, and owning a cell phone. Similarly, women who are married are less likely to have completed secondary or further education.

Variations in government effectiveness for women in Tanzania

In the section below, I examine the impact of gender equity in terms of access to the mobile ecosystem on knowledge diffusion and women's access to health care and education. Social justice perspectives presume health care and education to be public goods that should be provided for evenly to all citizens by an effective government.

The results from an analysis of the effect of residence, marriage, literacy, and cell phone ownership on whether women have drinking water on the premises of their homes are shown in Table 3.6. Women living in Zanzibar are 5.048 times as likely to have drinking water on the premises of their homes compared to those living in Pemba, Arusha, and all other (omitted) regions of Tanzania (excluding Kilimanjaro and Dar es Salaam). Those living in Kilimanjaro are 10.8 times as likely (relative to other omitted regions) and those in urban areas are 10.1 times as likely

Table 3.5 Determinants of secondary or further education: women aged 15–49, Tanzania (2015–2016)

Variables in the equation	*B*	*S.E.*	*Wald*	*df*	*Sig.*	*Exp (B)*	*95% C.I. for Exp(B)*	
							Lower	*Upper*
Zanzibar	1.171	0.139	70.6	1	0.000	3.227	2.455	4.241
Kilimanjaro	0.436	0.141	9.6	1	0.002	1.547	1.174	2.038
Dar es Salaam	0.383	0.072	28.2	1	0.000	1.466	1.273	1.689
Woman is currently married	–0.359	0.056	41.7	1	0.000	0.698	0.626	0.779
Urban residence	0.918	0.063	214	1	0.000	2.505	2.215	2.832
Female respondent owns cell phone	1.9	0.08	570	1	0.000	6.683	5.718	7.811
Constant	–3.606	0.078	2141	1	0.000	0.027		

Source: Zerai's analysis of TDHS (2015–2016).

Note: B = coefficient; S.E. = standard error; Wald = Wald statistic; df = degrees of freedom; Sig. = significance level; Exp(B) = odds ratio; C.I. = confidence interval.

Table 3.6 Determinants of access to drinking water on premises: women aged 15–49, Tanzania (2015–2016)

Variables in the equation	*B*	*S.E.*	*Wald*	*df*	*Sig.*	*Exp (B)*	*95% C.I. for Exp(B)*	
							Lower	*Upper*
Zanzibar	1.619	0.14	134.406	1	0.000	5.048	3.839	6.637
Kilimanjaro	2.381	0.132	327.228	1	0.000	10.815	8.356	13.998
Dar es Salaam	–0.05	0.065	0.595	1	0.441	0.951	0.838	1.08
Woman is currently married	–0.016	0.047	0.117	1	0.732	0.984	0.898	1.079
Literacy—reads passage easily	0.425	0.059	51.752	1	0.000	1.53	1.363	1.718
Urban residence	2.314	0.053	1,936.595	1	0.000	10.111	9.121	11.209
Female respondent owns cell phone	0.552	0.05	119.719	1	0.000	1.737	1.574	1.918
Constant	–2.719	0.063	1,856.537	1	0.000	0.066		

Source: Zerai's analysis of TDHS (2015–2016).

Note: B = coefficient; S.E. = standard error; Wald = Wald statistic; df = degrees of freedom; Sig. = significance level; Exp(B) = odds ratio; C.I. = confidence interval.

Table 3.7 Determinants of access to improved sanitation: women aged 15–49, Tanzania (2015–2016)

Variables in the equation	*B*	*S.E.*	*Wald*	*df*	*Sig.*	*Exp (B)*	*95% C.I. for Exp(B)*	
							Lower	*Upper*
Zanzibar	2.488	0.141	311.41	1	0.000	12.041	9.133	15.874
Kilimanjaro	1.441	0.117	152.646	1	0.000	4.224	3.361	5.309
Dar es Salaam	0.548	0.065	71.456	1	0.000	1.729	1.523	1.963
Woman is currently married	–0.068	0.048	2.054	1	0.152	0.934	0.851	1.025
Literacy—reads passage easily	0.794	0.067	141.75	1	0.000	2.211	1.94	2.52
Urban residence	1.332	0.054	617.223	1	0.000	3.79	3.412	4.21
Female respondent owns cell phone	0.679	0.053	163.692	1	0.000	1.971	1.776	2.187
Constant	–3.081	0.071	1,881.828	1	0.000	0.046		

Source: Zerai's analysis of TDHS (2015–2016).

Note: B = coefficient; S.E. = standard error; Wald = Wald statistic; df = degrees of freedom; Sig. = significance level; Exp(B) = odds ratio; C.I. = confi dence interval.

(relative to rural areas) to have drinking water on their premises. There is no significant difference for women living in Dar es Salaam, beyond the effects of living in urban areas. The same holds true for married women. If a woman owns a cell phone, she is 1.57 times more likely to have drinking water on her premises, net of other affects. All statistically significant results are at the $p<.001$ level.

Improved unshared toilets are related to living in Zanzibar, Kilimanjaro, and urban areas, with the strongest effect for those living in Zanzibar, where women are 12 times more likely to have an unshared improved toilet compared to women in other regions of the Tanzania. Literacy and owning a cell phone are both associated with women living in households with unshared improved toilet facilities. These highly statistically significant results can be found in Table 3.7.

Barriers to women's health care

Tables 3.8 and 3.9 show results of logistic regression analysis of financial and distance barriers to women obtaining their own health care. The

Table 3.8 Determinants of whether cost is a barrier to respondent obtaining her own health care: women aged 15–49, Tanzania (2015–2016)

Variables in the equation	*B*	*S.E.*	*Wald*	*df*	*Sig.*	*Exp (B)*	*95% C.I. for Exp(B)*	
							Lower	*Upper*
Zanzibar	–0.787	0.14	31.494	1	0.000	0.455	0.346	0.599
Kilimanjaro	–1.327	0.147	81.003	1	0.000	0.265	0.199	0.354
Dar es Salaam	0.334	0.066	25.636	1	0.000	1.397	1.227	1.59
Woman is currently married	0.041	0.037	1.256	1	0.262	1.042	0.97	1.119
Literacy—reads passage easily	–0.303	0.042	53.315	1	0.000	0.738	0.681	0.801
Urban residence	–0.688	0.046	222.45	1	0.000	0.503	0.459	0.55
Female respondent owns cell phone	–0.168	0.039	18.247	1	0.000	0.846	0.783	0.913
Constant	0.221	0.04	30.266	1	0.000	1.247		

Source: Zerai's analysis of TDHS (2015–2016).

Note: B = coefficient; S.E. = standard error; Wald = Wald statistic; df = degrees of freedom; Sig. = significance level; Exp(B) = odds ratio; C.I. = confidence interval.

Table 3.9 Determinants of whether distance is a barrier to respondent obtaining her own health care: women aged 15–49, Tanzania (2015–2016)

Variables in the equation	*B*	*S.E.*	*Wald*	*df*	*Sig.*	*Exp(B)*	*95% C.I. for Exp(B)*	
							Lower	*Upper*
Zanzibar	–0.744	0.13	32.649	1	0.000	0.475	0.368	0.613
Kilimanjaro	–0.758	0.118	41.25	1	0.000	0.469	0.372	0.591
Dar es Salaam	–0.084	0.064	1.712	1	0.191	0.92	0.811	1.043
Woman is currently married	–0.194	0.036	28.965	1	0.000	0.823	0.767	0.884
Literacy—reads passage easily	–0.476	0.042	130.016	1	0.000	0.622	0.573	0.674
Urban residence	–0.303	0.044	47.63	1	0.000	0.739	0.678	0.805
Female respondent owns cell phone	–0.251	0.039	42.466	1	0.000	0.778	0.721	0.839
Constant	0.697	0.041	290.067	1	0.000	2.007		

Source: Zerai's analysis of TDHS (2015–2016).

Note: B = coefficient; S.E. = standard error; Wald = Wald statistic; df = degrees of freedom; Sig. = significance level; Exp(B) = odds ratio; C.I. = confi dence interval.

results are similar in the two tables. Living in Zanzibar, Kilimanjaro, and urban areas is associated with women who do not have financial or transportation barriers to obtaining their own health care. However, living in Dar es Salaam has variable effects. Women living there are 1.397 times more likely to have financial barriers compared to women in other (omitted) regions of Tanzania, whereas women living there are less likely to have distance barriers to obtaining health care. Women who are literate and own cell phones are less likely to have financial or distance barriers to obtaining health care.

In Table 3.10 I examine the potential impact of gender equity in women's access to the mobile ecosystem on women's perspectives on intimate partner violence.

Attitudes about IPV vary considerably throughout Tanzania. In urban areas (including Dar es Salaam), Zanzibar, and Kilimanjaro, women are not likely to agree that intimate partner violence is justified if a wife refuses to have sex. Women who are literate and who own cell phones are also not likely to agree that IPV is justified in that case. However, Table 3.10 shows that married women are likely to agree that IPV is justified if a wife refuses sex with her husband.

Table 3.10 Determinants of respondent's disagreement that IPV is justified if wife refuses to have sex: women aged 15–49, Tanzania (2015–2016)

Variables in the equation	*B*	*S.E.*	*Wald*	*df*	*Sig.*	*Exp (B)*	*95% C.I. for Exp(B)*	
							Lower	*Upper*
Zanzibar	0.482	0.141	11.71	1	0.001	1.62	1.229	2.135
Kilimanjaro	0.789	0.139	32.066	1	0.000	2.2	1.675	2.891
Dar es Salaam	0.328	0.075	19.09	1	0.000	1.388	1.198	1.609
Literacy—reads passage easily	0.341	0.042	64.902	1	0.000	1.406	1.294	1.528
Woman is currently married	–0.133	0.038	12.257	1	0.000	0.876	0.813	0.943
Urban residence	0.389	0.047	67.457	1	0.000	1.476	1.345	1.62
Female respondent owns cell phone	0.272	0.041	44.711	1	0.000	1.312	1.212	1.421
Constant	0.159	0.041	15.172	1	0.000	1.172		

Source: Zerai's analysis of TDHS (2015–2016).

Note: B = coefficient; S.E. = standard error; Wald = Wald statistic; df = degrees offreedom; Sig. = significance level; Exp(B) = odds ratio; C.I. = confidence interval.

Summary

In Chapter 3, we have discovered that women aged 15–49 who live in Zanzibar, Kilimanjaro, Dar es Salaam, and urban areas are more likely to have their own cell phones. Cell phone ownership is associated with women's literacy and access to important development resources such as clean water and safe sanitation, as well as better access to their own health care because money and distance are not likely to be problems. In Chapter 4, I will examine these issues in Malawi.

Notes

1 *Region* is entered into analyses in Tables 3.3–3.10 as four categories: Zanzibar, Kilimanjaro, Dar es Salaam, and a fourth *omitted* category to which the first 3 regions are compared. The omitted category includes Pemba, Arusha, and all other (omitted) regions of Tanzania (excluding Zanzibar, Kilimanjaro, Dar es Salaam).

2 To ease readability of chapters, details of reports of statistical analyses are streamlined in the text. The unit of analysis for all logistic regression analyses in chapters 3 and 4 are women aged 15–49 in the Demographic and Health Survey samples.

3 The literacy variable is operationalized utilizing an ordinal scale in the following way. A woman is literate if she has either completed secondary education or reads a passage easily. If she has not completed secondary education and cannot read a passage at all, she is not considered to be literate. If she has not completed secondary education and reads a passage with difficulty, she is considered partially literate.

4 ICT, women's status, and governance in Malawi, 2010 and 2015–2016

In a Malawi case study on the interplay of state policy, family structure, and land on women and children's wellbeing, Professor Linda Semu (2011, p. 57) identifies wellbeing from the perspectives of women and men in Zomba District.

> Results show that villagers typically define well-off households as those having: good-quality housing, food lasting the whole year, livestock, a bicycle, a personable character and the ability to purchase farming inputs (fertilizer and hybrid seed), and routine daily needs. Less well-off households are defined as having characteristics that are the opposite of those of the well-off households, including inadequate land, reliance on piece-work, and being the most negatively affected by rising commodity prices.

Though these definitions do not directly address gender, gender features prominently in Professor Semu's academic work. A distinction of Malawi compared to the other two countries under close examination in this manuscript is that the majority of individuals trace their lineage through their mother's line. Semu (2005:5) notes:

> The matrilineal system acts as a leveling mechanism that does not necessarily give women an unfair advantage over men, but gives both single and married women the chance to access land. However, matrilineal women's autonomy has become increasingly eroded in the face of diminishing land size, failures in state policy, unreliable climatic conditions, and the devastation brought about by HIV/AIDS.

The main external factors to exert a positive influence on gender mainstreaming have been international pressure and the role of NGOs. But Semu and her colleagues argue that these affects have largely been just for show and short-lived when it comes to governance. For example:

> The year 1982 is the only year in the history of Malawi when a large number of women were serving together in Parliament … It was important for Banda to nominate a large number of women to parliament at that election, since preparations were under way for the Women's Conference in Nairobi. Nominating large numbers of women presented the MCP government, and in particular President Banda, in a positive light regarding women's empowerment. By 1987, two years after the women's conference, only one woman was in parliament.
>
> (Semu 2002, p. 86)

Interest convergence from critical race theory (Bell 1980) is a relevant concept in this case. Backing for women's issues, and the promotion of women into leadership positions within the government, was supported by the male dominant government only to the extent that these promotions represented political capital for President Banda. This same problem held true when Malawi signed the United Nations Convention on the Elimination of All Forms of Discrimination against Women (CEDAW), and policy-makers and law-makers maintained that CEDAW was acceptable only to the extent that it did not interfere with "traditional" culture in Malawi (Semu 2002).

It is therefore extremely important to take into account both the factors that define wellbeing for Malawians and also to identify factors that specifically address women's status when we think about access to cell phones, knowledge diffusion, and good governance.

Cell phones in households

In Malawi, 53.7% of households had at least one cell phone in 2015–2016, compared to 39% in 2010 and 4.7% in 2004, according to the Malawi Demographic and Health Survey (MWDHS, 2004, 2010, 2015–2016). Given the robust ICT network in Malawi, below I examine female scholars' and activists' electronic footprint to see the extent to which their ideas engage with local and national governance. As in previous chapters, I foreground women scholars' and activists' discussions of gender equity. There are vast differences in the lived experiences of educated and economically well-endowed women compared to their less-educated and impoverished counterparts. So, I examine regional differences among women, as well as differences according to socio-economic status, and other measures of class status in Malawi. Looking at household-level data in the MWDHS, the percentage of child-bearing-aged women's households with landlines has remained constant over the past decade. In 2015–2016 it was 2.1% (and 2% in 2004 and 2010). In 2015–2016, we have information available on households with computers showing that 3.5% of households possessed them.

Table 4.1 Crosstabulation table of region of residence and native language of respondent: Malawi (2015–2016) (count and percentage)

	Native language of respondent				*Total n (100%)*
	English	*Chichewa*	*Tumbuka*	*Other*	
Northern region	1	258	2,418	162	2,839
	0%	9.1%	85.2%	5.7%	
Central region	79	10,258	104	88	10,529
	0.8%	97.4%	1%	0.8%	
Southern region	12	9,695	28	1,459	11,194
	0.1%	86.6%	0.3%	13%	
Total	92	20,211	2,550	1,709	24,562
	0.4%	82.3%	10.4%	7%	100%

Source: Zerai's analysis of MWDHS (2015–2016).

As is the case with Zimbabwe and Tanzania, a distinguishing feature of the MWDHS data is that we have information on the percentage of women who own cell phones. In 2015–2016, one-third (33%) of women owned them, compared to over half of men (52%). This shows that cell phones in households largely belong to men.

Financial resources provide another sense of women's status vis-à-vis men. In relation to bank accounts, 17% of men and 10% of women had them in 2015–2016. But for those who have cell phones, women were more likely to use cell phones for financial transactions. Though 33% of women own phones, 27% use them for bank transactions, compared to only 30% of men.

Language data are available in the Malawi DHS! This is exciting, because many of the African DHS datasets have stripped out native-language data. As shown in Table 4.1, the native languages captured in the MWDHS are Chichewa (82% of respondents), Tumbuka (10.4%), English (less than 1% or only 92 out of 24,562 respondents), and an "Other" category (7%).[1] The languages do not line up perfectly by region in Malawi, but there are some pretty strong correlations (which are highly statistically significant, according to chi-square results).

In 2015–2016, the majority native language of respondents in the Central and Southern regions (97% and 87%, respectively) was Chichewa, whereas the majority of respondents spoke Tumbuka in the northern region (85.2%). Breaking this down further into rural and urban areas, the rural southern region had 15.1% of native speakers of "other" languages (see Table 4.2, which yields highly statistically significant chi-square results indicating that the native language of respondents and their region/rural or urban areas of residence are not

Table 4.2 Crosstabulation table of region/rural of residence and native language of respondent: Malawi (2015–2016) (count and percentage)

	Native language of respondent				*Total n (100%)*
	English	*Chichewa*	*Tumbuka*	*Other*	
Urban Northern region*	0	116	385	12	513
	0%	22.6%	75%	2.3%	
Urban Central region	57	1,921	13	24	2,015
	2.8%	95.3%	0.6%	1.2%	
Urban Southern region	2	1,896	4	65	1,967
	0.1%	96.4%	0.2%	3.3%	
Rural Northern region	0	142	2,033	150	2,325
	0%	6.1%	87.4%	6.5%	
Rural Central region	22	8,336	91	65	8,514
	0.3%	97.9%	1.1%	0.8%	
Rural Southern region**	10	7,798	24	1,395	9,227
	0.1%	84.5%	0.3%	15.1%	
Total	91	20,209	2,550	1,711	24,561
	0.3%	82.2%	10.3%	6.9%	100.00%

Source: Zerai's analysis of MWDHS (2015–2016).

* In the MWDHS 2010, Tumbuka native-language speakers comprised 54% of residents in the urban northern region, and "other" languages were 43%.

** In MWDHS 2010, Yao native-language speakers comprised 20% of residents in the rural southern region, and "other" languages comprised 18%.

independent factors). So, I created an ethnicity category that captured ethno-linguistic differences between three groups: native Chichewa speakers (82%), native Tumbuku speakers (10%), and native "Other" language speakers (7%). I also grouped regions by dominant ethno-linguistic groups: the Chewa region, the Tumbuku region, and a "mixed/other" region, the rural Southern region, that includes 15% "Other" language speakers, and 85% Chichewa.

Summary of literature from women and gender-sensitive scholars in and of Malawi

Some of the main themes in the literature that are incorporated into the conceptual framework below include: (1) *gender*, including matrilineal family structure and women's rights (Semu 2002, 2005, and

2011), gender-based programs (Semu 2002, 2011; Kowalczyk et al. 2014), exploitation of gender by politicians (Semu 2011; Henderson and Gilman 2004), and family wellbeing (Semu 2011); (2) *mobile technology* (Chima and Chikasanda 2013; Chawinga, Winner, and Ngwira 2013); (3) *mobile technology and health* (Huggins and Valverde 2018; Crawford et al. 2014); (4) *knowledge diffusion and health* (Meekers et al. 2007; Kaufman et al. 2014; Chaputula and Mutula 2018; Chikasanda et al. 2012; Chikasanda 2015); and (5) a number of other topics more squarely focused on health, gender, and food insecurity, sanitation, or commerce that do not directly address ICT (Wiza et al. 2017; Ntata et al. 2008; Gine and Yang 2009; Chunga et al. 2016.; Kakota et al. 2011).

These studies point to the fact that in Malawi, gender has often been used for the purpose of political expediency, where gender analysis or values promoting gender equity have been subsumed under androcentric versions of culture (Semu 2002). These findings align with the trend that gender-based programs often lack a systems approach (Kowalczyk et al. 2014) and that gender is often not considered in the creation of development technologies (Zerai and Morrow 2018). Studies of cell phone technology and its impact on health and knowledge diffusion are gaining traction. However, while internet banking is becoming more popular (Chima and Chikasand 2013), problems with slow connections and the dearth of available services impedes usage (Chawinga et al. 2015). Just as in other parts of the world, mobile messaging can more greatly increase positive behavioral change and attitudes in relation to health than voice messages in rural Malawi (Crawford et al. 2014). An analysis of a mobile nutrition program, referred to as "mNutrition," finds this technology to be effective (Huggins and Valverde 2018). And Crawford et al. (2014) found that "mobile SMS health messages had higher successful delivery and led to higher intended or actual behavior change among subscribers than voice messages." Furthermore:

> [O]verall, 99% of survey respondents reported trusting messages they received, and about 75% of respondents recalled the last message they received and learned something new. Almost 75% of respondents reported that they had already changed or intended to change their behavior based on received messages. Intended or actual behavior change was significantly higher among pushed SMS enrollees than among pushed or retrieved voice messaging enrollees.
>
> (Crawford et al. 2014, p. 35)

However, other modes of delivery, such as radio communications, were also found to be effective. A study by Kaufman et al. (2014) found that risky sexual behavior changed as a result of a popular radio show.

The conceptual framework

The conceptual framework is similar to that found in Chapter 3 (see Figure 4.1). First, I examine the impact of woman's literacy, country cleavages, and marriage on women's cell phone ownership. Then I examine the impact of these factors, in addition to women's cell phone ownership, on various outcomes relevant to women. These include women's views on wife beating (i.e. intimate partner violence) and on outcomes in individual women's lives that signal general government effectiveness, including higher equity in access to water/sanitation, lack of barriers to health care for the respondent (distance and cost), and educational attainment.

Linda Semu (2002, 2005, 2011) points to the importance of ethnic and regional differences in Malawi. And she notes that: "Malawian women now have to contend with political party affiliation that tends to follow ethnic affiliation" (2002, p. 87). I therefore incorporate them into my conceptual framework. For the in-country cleavage in Malawi, I include the woman's native language in some models to get a sense of whether ethno-linguistic differences are relevant, and in other models I include the dominant ethnicity in the woman's region of residence, as explained above.

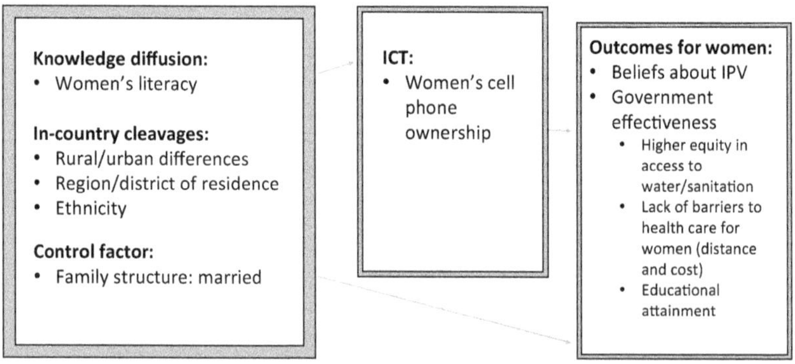

Figure 4.1 Ownership of cell phones and individual outcomes for women aged 15–49, Malawi

Women's ownership of cell phones in various regions of Malawi

Taking a look at cell phone ownership within rural/urban areas and region, the results are as expected for the most part. As shown in Table 4.3, 33% of all women respondents in the Malawi DHS own a cell phone. And 66% of women from urban areas of the Northern region as well as 66% in the urban Central region own a cell phone. The smallest percentage of women who own a cell phone reside in the rural areas of the Central region (23%), compared to other regions, taking rural/urban areas into account. Twice as many women in urban regions own cell phones compared to rural areas within the same region. The chi-square statistic indicates that these results are highly statistically significant.

Table 4.4 provides descriptive statistics for variables in the logistic regression analyses. The average age of respondents is 28 in the MWDHS, and 62% are married. Women's literacy and educational attainment provide a sense of vectors for knowledge diffusion. The majority of women (64%) either completed secondary or higher levels of education or read a passage in English, Chichewa or Tumbuka easily, while 9% read the passage with difficulty. The lack of ability to read the passage, blindness, or the lack of a card printed in the respondent's language accounts for 26% of women (the omitted category).

The percentage of women with incomplete primary or no formal education in Malawi is high (66%), especially given the literacy rate. Only 10% of women have completed secondary or further education.

In relation to access to water, sanitation, and health care, 17% had water on their premises, 54% had unshared improved toilets, and a little over half of women aged 15–49 found a lack of money for treatment

Table 4.3 Number and percentage of women aged 15–49 who own cell phones, by region and rural/urban area in Malawi (2015–2016)

Region	*Number*	*Percentage*
Urban Northern region	340	66.3%
Urban Central region	1,336	66.3%
Urban Southern region	1,195	60.8%
Rural Northern region	875	37.6%
Rural Central region	1,980	23.3%
Rural Southern region	2,340	25.4%
Total	8,066	32.8%

Source: Zerai's analysis of MWDHS (2015–2016).

Note: n=24,560

Table 4.4 Descriptive statistics, unit of analysis: women aged 15–49, Malawi (2015–2016)

Variable	*Descriptive statistics*		
	Number	*Mean or proportion*	*Standard deviation*
ICT			
Household has landline telephone	24,311	0.03	0.16
Household has cell phone			
Female respondent owns a cell phone	24,562	0.33	0.47
Female respondent uses own cell phone for financial transactions	8,066	0.27	0.45
Knowledge diffusion			
Literacy—reads passage easily	24,562	0.64	0.48
Literacy—reads passage with difficulty	24,562	0.09	0.28
Female respondent, highest educational level, secondary or further	24,562	0.10	0.29
Female respondent, incomplete primary or no formal education	24,562	0.66	0.47
Government effectiveness/development resources availability			
Drinking water on premises	24,562	0.17	0.38
Unshared improved toilets	24,562	0.54	0.50
Barriers to woman's health care: lack of money for treatment a big problem	24,562	0.56	0.50
Barriers to woman's health care: distance to facility a big problem	24,562	0.53	0.50
Control variables			
Urban residence	24,562	0.18	0.39
Age of female respondent (15–49)	24,562	28.11	9.26
Resident is a visitor (versus usual resident)	24,562	0.01	0.10
Respondent is currently married	24,562	0.62	0.49
Respondent disagrees wife beating is justified if woman refuses to have sex	24,562	0.91	0.29
Native language			
Chichewa	24,562	0.82	0.38
Tumbuka	24,562	0.10	0.31
Other	24,562	0.07	0.25

Source: Zerai's analysis of MWDHS (2015–2016).

Note: For all variables, the minimum value is 0 and the maximum is 1, unless otherwise indicated.

or distance to a health facility to be barriers to getting medical help for themselves. Most disagreed with the idea that "wife beating is justified" if woman refuses to have sex (91%).

Differential access to cell phones

Women living in Tumbuka-majority regions are more likely to own cell phones compared to those in the Southern rural region. And those living in Chewa-majority regions are less likely to own cell phones compared to the southern rural region. Women in urban areas are 4.65 times more likely to own a cell phone compared to women in rural areas, over and above the effect of other factors. Married women are 1.439 times more likely to own a cell phone compared to women who are not married. And women who have obtained secondary or further education or can read a passage easily in English, Chichewa or Tumbuka are 2.81 times more likely to own cell phones (see Table 4. 5).

Tables 4.6–4.8 provide logistic regression results for an analysis of native language, cell phone ownership, marriage, rural/urban residence,

Table 4.5 Determinants of ownership of cell phones: women aged 15–49, Malawi (2015–2016)

Variables in the equation	*B*	*S.E.*	*Wald*	*df*	*Sig.*	*Exp (B)*	*95% C.I. for Exp(B)*	
							Lower	*Upper*
Woman lives in Tumbuka-majority region	0.381	0.05	63.33	1	0.000	1.464	1.333	1.609
Woman lives in Chewa-majority region	–0.12	0.04	11.5	1	0.001	0.887	0.828	0.951
Woman is currently married	0.364	0.03	139.4	1	0.000	1.439	1.355	1.529
Literacy—reads passage easily	1.031	0.03	936.5	1	0.000	2.805	2.625	2.996
Urban residence	1.537	0.04	1477	1	0.000	4.65	4.299	5.029
Constant	–1.956	0.04	2348	1	0.000	0.141		

Source: Zerai's analysis of MWDHS (2015–2016).

Note: B = coefficient; S.E. = standard error; Wald = Wald statistic; df = degrees of freedom; Sig. = significance level; Exp(B) = odds ratio; C.I. = confidence interval.

Table 4.6 Determinants of secondary or further education: women aged 15–49, Malawi (2015–2016)

Variables in the equation	*B*	*S.E.*	*Wald*	*df*	*Sig.*	*Exp(B)*	*95% C.I. for Exp(B)*	
							Lower	*Upper*
Woman's native language: Tumbuka	0.014	0.13	0.01	1	0.916	1.014	0.786	1.308
Woman's native language: Chichewa	0.134	0.11	1.5	1	0.221	1.143	0.923	1.416
Woman is currently married	–0.343	0.05	47	1	0.000	0.709	0.643	0.783
Urban residence	1.592	0.05	1000	1	0.000	4.912	4.451	5.422
Female respondent owns cell phone	2.351	0.06	1419	1	0.000	10.493	9.285	11.858
Constant	–4.082	0.12	1181	1	0.000	0.017		

Source: Zerai's analysis of MWDHS (2015–2016).

Note: B = coefficient; S.E. = standard error; Wald = Wald statistic; df = degrees of freedom; Sig. = significance level; Exp(B) = odds ratio; C.I. = confidence interval.

Table 4.7 Determinants of incomplete primary or no formal education: women aged 15–49, Malawi (2015–2016)

Variables in the equation	*B*	*S.E.*	*Wald*	*df*	*Sig.*	*Exp(B)*	*95% C.I. for Exp(B)*	
							Lower	*Upper*
Woman's native language: Tumbuka	–0.758	0.07	109	1	0.000	0.468	0.406	0.54
Woman's native language: Chichewa	–0.15	0.06	6.09	1	0.014	0.861	0.764	0.97
Woman is currently married	0.409	0.03	174	1	0.000	1.505	1.417	1.6
Urban residence	–1.479	0.04	1469	1	0.000	0.228	0.211	0.246

Variables in the equation	*B*	*S.E.*	*Wald*	*df*	*Sig.*	*Exp(B)*	*95% C.I. for Exp(B)*	
							Lower	*Upper*
Female respondent owns cell phone	–1.335	0.03	1806	1	0.000	0.263	0.247	0.28
Constant	1.415	0.06	524	1	0.000	4.115		

Source: Zerai's analysis of MWDHS (2015–2016).

Notes:
In Tables 4.6 and 4.7, the respondent's native language is examined (and not dominant ethnicity in the region).

B = coefficient; S.E. = standard error; Wald = Wald statistic; df = degrees of freedom; Sig. = significance level; Exp(B) = odds ratio; C.I. = confidence interval.

Table 4.8 ICT and knowledge diffusion: determinants of whether woman (aged 15–49) reads passage easily, Malawi (2015–2016)

Variables in the equation	*B*	*S.E.*	*Wald*	*df*	*Sig.*	*Exp(B)*	*95% C.I. for Exp(B)*	
							Lower	*Upper*
Woman's native language: Tumbuka	0.614	0.068	82.5	1	0.000	1.847	1.618	2.109
Woman's native language: Chichewa	0.204	0.052	15.412	1	0.000	1.226	1.107	1.357
Woman is currently married	–0.505	0.029	297.425	1	0.000	0.603	0.57	0.639
Urban residence	0.982	0.046	459.676	1	0.000	2.669	2.44	2.92
Female respondent owns cell phone	1.029	0.034	934.205	1	0.000	2.799	2.62	2.99
Constant	0.212	0.053	16.008	1	0.000	1.237		

Source: Zerai's analysis of MWDHS (2015–2016).

Notes:
Respondent's native language is examined (and not the dominant ethnicity in the region).

B = coefficient; S.E. = standard error; Wald = Wald statistic; df = degrees of freedom; Sig. = significance level; Exp(B) = odds ratio; C.I. = confidence interval.

Table 4.9 Determinants of access to drinking water on premises: women aged 15–49, Malawi (2015–2016)

Variables in the equation	*B*	*S.E.*	*Wald*	*df*	*Sig.*	*Exp (B)*	*95% C.I. for Exp(B)*	
							Lower	*Upper*
Woman lives in Tumbuka-majority region	0.44	0.067	42.874	1	0.000	1.553	1.361	1.772
Woman lives in Chewa-majority region	–0.111	0.056	3.901	1	0.048	0.895	0.802	0.999
Woman is currently married	–0.184	0.041	20.056	1	0.000	0.832	0.767	0.902
Literacy—reads passage easily	0.407	0.049	69.614	1	0.000	1.503	1.365	1.653
Urban residence	2.543	0.05	2,583.585	1	0.000	12.712	11.525	14.021
Female respondent owns cell phone	0.806	0.042	369.469	1	0.000	2.24	2.063	2.432
Constant	–2.872	0.058	2,478.823	1	0.000	0.057		

Source: Zerai's analysis of MWDHS (2015–2016).

Note: B = coefficient; S.E. = standard error; Wald = Wald statistic; df = degrees of freedom; Sig. = significance level; Exp(B) = odds ratio; C.I. = confidence interval.

and knowledge diffusion (dependent variable). The knowledge diffusion dependent variable is measured as secondary or further education, incomplete primary or no formal education, and literacy among women aged 15–49 in 2015–2016.

The results show that secondary or higher educational attainment is positively related to urban residence and cell phone ownership, and is negatively related to marriage. These results are significant at $p<.001$ (see Table 4.6). Furthermore, the results show that no formal education or incomplete primary educational attainment is negatively related to Tumbuku and Chichewa native languages, urban residence, and cell phone ownership. And educational attainment is positively related to marriage (see Table 4.7). Finally, Table 4.8 provides consistent results in that owners of cell phones are 2.8 times more likely to easily read a

Table 4.10 Determinants of access to unshared flush toilets: women aged 15–49, Malawi (2015–2016)

Variables in the equation	*B*	*S.E.*	*Wald*	*df*	*Sig.*	*Exp (B)*	*95% C.I. for Exp(B)*	
							Lower	*Upper*
Woman lives in Tumbuka-majority region	0.062	0.044	1.961	1	0.1610	1.064	0.976	1.16
Woman lives in Chewa-majority region	0.076	0.03	6.489	1	0.0110	1.079	1.018	1.145
Woman is currently married	–0.107	0.027	16.038	1	0.000	0.898	0.852	0.947
Literacy—reads passage easily	0.158	0.028	31.944	1	0.000	1.172	1.109	1.238
Urban residence	–0.428	0.038	125.232	1	0.000	0.652	0.604	0.702
Female respondent owns cell phone	0.367	0.03	150.977	1	0.000	1.443	1.361	1.53
Constant	0.02	0.032	0.389	1	0.5330	1.02		

Source: Zerai's analysis of MWDHS (2015–2016).

Note: B = coefficient; S.E. = standard error; Wald = Wald statistic; df = degrees of freedom; Sig. = significance level; Exp(B) = odds ratio; C.I. = confidence interval.

passage compared to counterparts who do not own a cell phone. And women whose native languages are Tumbuku or Chichewa are also more likely to read a passage easily. Marriage is negatively related to literacy[2], and women residing in urban areas are 2.67 times more likely to easily read a passage. All the results in Table 4.8 are highly statistically significant ($p<.001$).

Tables 4.9 and 4.10 provide results of analyses of impact of cell phone ownership and knowledge diffusion on differential access to safe drinking water and improved/safe sanitation (measured as an unshared flush toilet on premises) among women aged 15–49 in Malawi in 2015–2016. Marriage is negatively related to clean water and safe sanitation. Literacy and cell phone ownership are both positively related to clean water and safe sanitation. However, women living in a

Tumbuka-majority region are more likely to have access to clean water. But living in Tumuku-majority regions has no effect on unshared flush toilets at home. And the opposite is true for women in Chewa-dominated regions, who are more likely to have access to sanitation and less likely to have access to proximate clean drinking water. This may all seem counterintuitive, but it is actually good news that statistically significant differences do not show a clear pattern of preference for one region over another in relation to the ethno-linguistic groups that dominate regions.

Barriers to obtaining women's health care: women aged 15–49, Malawi (2015–2016)

Tables 4.11 and 4.12 focus on financial and distance barriers to women obtaining health care for themselves. The way in which these are

Table 4.11 Determinants of whether cost is a barrier to woman obtaining her own medical care: women aged 15–49, Malawi (2015–2016)

Variables in the equation	*B*	*S.E.*	*Wald*	*df*	*Sig.*	*Exp (B)*	*95% C.I. for Exp(B)*	
							Lower	*Upper*
Woman lives in Tumbuka-majority region	–1.031	0.047	472.002	1	0.000	0.357	0.325	0.391
Woman lives in Chewa-majority region	0.178	0.031	33.752	1	0.000	1.195	1.125	1.269
Woman is currently married	–0.099	0.028	13.041	1	0.000	0.905	0.858	0.956
Literacy—reads passage easily	–0.197	0.029	46.788	1	0.000	0.821	0.776	0.869
Urban residence	–0.731	0.04	339.809	1	0.000	0.481	0.445	0.52
Female respondent owns cell phone	–0.364	0.03	146.081	1	0.000	0.695	0.655	0.737
Constant	0.576	0.033	311.22	1	0.000	1.778		

Source: Zerai's analysis of MWDHS (2015–2016).

Note: B = coefficient; S.E. = standard error; Wald = Wald statistic; df = degrees of freedom; Sig. = significance level; Exp(B) = odds ratio; C.I. = confi dence interval.

Table 4.12 Determinants of whether distance is a barrier to respondent obtaining her own medical care: women aged 15–49, Malawi (2015–2016)

Variables in the equation	*B*	*S.E.*	*Wald*	*df*	*Sig.*	*Exp (B)*	*95% C.I. for Exp(B)*	
							Lower	*Upper*
Woman lives in Tumbuka-majority region	–0.482	0.045	116.24	1	0.000	0.617	0.566	0.674
Woman lives in Chewa-majority region	0.156	0.031	25.568	1	0.000	1.169	1.1	1.242
Woman is currently married	–0.036	0.027	1.67	1	0.196	0.965	0.914	1.019
Literacy—reads passage easily	–0.162	0.029	31.496	1	0.000	0.851	0.804	0.9
Urban residence	–0.973	0.04	606.021	1	0.000	0.378	0.35	0.408
Female respondent owns cell phone	–0.289	0.03	93.408	1	0.000	0.749	0.706	0.794
Constant	0.604	0.033	341.195	1	0.000	1.829		

Source: Zerai's analysis of MWDHS (2015–2016).

Note: B = coefficient; S.E. = standard error; Wald = Wald statistic; df = degrees of freedom; Sig. = significance level; Exp(B) = odds ratio; C.I. = confidence interval.

measured in the Malawi DHS is whether the respondent indicates that in relation to "getting medical help for self," that either "distance to health facility" is a problem or "getting money needed for treatment" is a problem. The results for the two variables are almost identical, except that marriage is not significant in the case of distance. On the other hand, marriage is negatively related to financial barriers. These barriers are negatively related to living in Tumbuka-majority regions, literacy, urban residence, and women's cell phone ownership. Women in Chewa-majority regions are more likely to experience both types of barriers.

Views on domestic violence

An analysis of whether respondent does not agree that IPV is justified if the wife refuses to have sex, among women 15–49 in Malawi can be found in Table 4.13. Instead of a regional analysis, I have analyzed the impact of individual ethnicity, literacy, marriage, urban residence, and

Table 4.13 Determinants of respondent's disagreement that IPV is justified if wife refuses to have sex: women aged 15–49, Malawi (2015–2016)

Variables in the equation	*B*	*S.E.*	*Wald*	*df*	*Sig.*	*Exp (B)*	*95% C.I. for Exp(B)*	
							Lower	*Upper*
Woman's native language: Tumbuka*	–0.924	0.103	81.109	1	0.000	0.397	0.325	0.485
Woman's native language: Chichewa*	–0.117	0.09	1.685	1	0.194	0.889	0.745	1.062
Literacy—reads passage easily	0.325	0.047	48.276	1	0.000	1.384	1.263	1.516
Woman is currently married	0.138	0.046	8.896	1	0.003	1.147	1.048	1.256
Urban residence	0.627	0.078	65.235	1	0.000	1.871	1.607	2.178
Female respondent owns cell phone	0.359	0.055	42.339	1	0.000	1.431	1.285	1.595
Constant	2.047	0.094	473.493	1	0.000	7.741		

Source: Zerai's analysis of MWDHS (2015–2016).

* Individual women's ethno-linguistic groups are more relevant than regional groupings.

Note: B = coefficient; S.E. = standard error; Wald = Wald statistic; df = degrees of freedom; Sig. = significance level; Exp(B) = odds ratio; C.I. = confidence interval.

cell phone ownership on views concerning domestic violence. Women whose native language is Tumbuka are more likely to agree that IPV is justified if a wife refuses to have sex with her husband, and women who read passages easily, are married, from urban areas, and who own cell phones are more likely to disagree.

Summary and lessons learned

The results show that ethnicity does not seem to be an overwhelming factor in Malawi. There is no evidence of favoritism. While development resources do not favor certain regions, the literature points to gender power imbalances, as shown by married women's views concerning wife beating (i.e. IPV), low levels of women's secondary or further education or completed primary education, and that if a household owns a cell phone, it is more likely to be owned by male members of that household. However, women's literacy levels are relatively high given their low formal educational attainment. Evidence of government effectiveness such as women's access to health care and development resources such as clean water and improved sanitation show that, by and large, women with better access to cell phones are more likely to benefit from these.

In Chapters 2, 3, and 4, I examined the literature from female scholars in Zimbabwe, Tanzania, and Malawi to build a new conceptual model for understanding the impact of ICT on good governance. Finally, in Chapter 5, I articulate new hypotheses that take the literature from female scholars in all three countries and beyond into account including intellectual works from other African women and gender-conscious scholars. I also analyze the new model that includes gender equity as a part and parcel of good governance.

Notes

1 The Yao language is included in the 2010 Malawi DHS. Then, 18% of residents spoke "Other" native languages, 7% of respondents spoke Yao, and 12% spoke Tumbuka. Yao is excluded as a category from the 2015–16 MWDHS.

2 As in chapter 3, the literacy variable is operationalized utilizing an ordinal scale in the following way. A woman is literate if she has either completed secondary education or reads a passage easily. If she has not completed secondary education and cannot read a passage at all, she is not considered to be literate. If she has not completed secondary education and reads a passage with difficulty, she is considered partially literate.

5 ICT, diffusion of knowledge to women, gender-inclusive governance, and impacts on women's lives in three African nations

In this chapter, I merge DHS data with World Bank and African Development Bank Group governance indicators in order to test for lag effects on the availability of cell phone technology in women's households. I also examine their contribution to outcomes that are important to women that should result from effective governance. Inspired by Asongu and Nwachukwu's 2016 analysis of 2000–2012 World Bank development and governance indicators, I examine whether cell phones contribute to the relative diffusion of knowledge in Zimbabwe, Tanzania, and Malawi. I also examine whether political, economic, and institutional governance in these countries has a positive effect on women's lives. Further, I examine "better governance" in a way that considers factors that are relevant to gender equity. Thus, I examine cell phones, knowledge diffusion, and people-centered governance to include measures of women's status.

If Asongu and Nwachukwu's theory that ICT and knowledge diffusion contribute to good governance is true, their measures of good governance should have a positive impact on the lives of individual women. As was shown in the previous chapters, the concerns of women include education and literacy, their economic wellbeing, political participation, access to good health care for their children and themselves, safety in marriage and in their communities, and a happy and peaceful home life. In short, if governance is good, it should also be good for women. I theorize that good governance should be related to gender parity in education, women's political participation, a lack of violence against women, and laws that prosecute intimate partner violence. It should also accompany social safety net protections in place for families in crisis, and equitable access to economic opportunities for women.

In the multilevel analyses shown below, I examine Zimbabwe, Tanzania, and Malawi together to see if ICT (women's ownership of cell phones), knowledge diffusion, and governance have predictive relationships with women's literacy, health care, and beliefs about

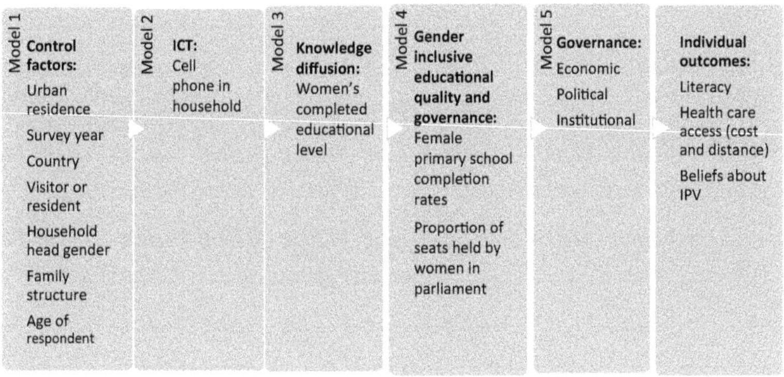

Figure 5.1 Models 1–5

intimate partner violence (see Figure 5.1). The multilevel analyses include country-level variables such as rural and urban residence, the proportion of women who completed primary school, the proportion of seats held by women in parliament, and aggregate indicators of six dimensions of governance from the World Bank's WGI from 2010 and 2014. Figure 5.2 shows the six dimensions of governance that are grouped into three categories: (a) political governance (political stability and absence of violence/terrorism and voice and accountability; (b) economic governance (regulatory quality and government effectiveness); and (c) institutional governance (control of corruption and rule of law). Finally, I examine country-level indicators of women's status in each category in 2010 and 2014 (Figure 5.3): for political governance, violence against women in the last 12 months and the proportion of women elected to parliament; for economic governance, women's equal pay, gender in nondiscrimination laws, and the availability of social safety nets; and for institutional governance, the existence of domestic violence legislation. My intervention is to ask whether variables that highlight women's access to the fruits of good governance improve the effectiveness of governance models for women. To test this, I compare models that include political, economic, and institutional governance predictors with models that include gender equity to see which best explain health care access (measured as whether money or distance is a barrier to health care), women's literacy, and beliefs about intimate partner violence. Notice that these dependent variables are measured at the individual level. I also examine specific directly relevant factors related to political, economic, and institutional governance indicators to see which have an impact on outcomes that are important to women. These include safety net coverage, gender parity

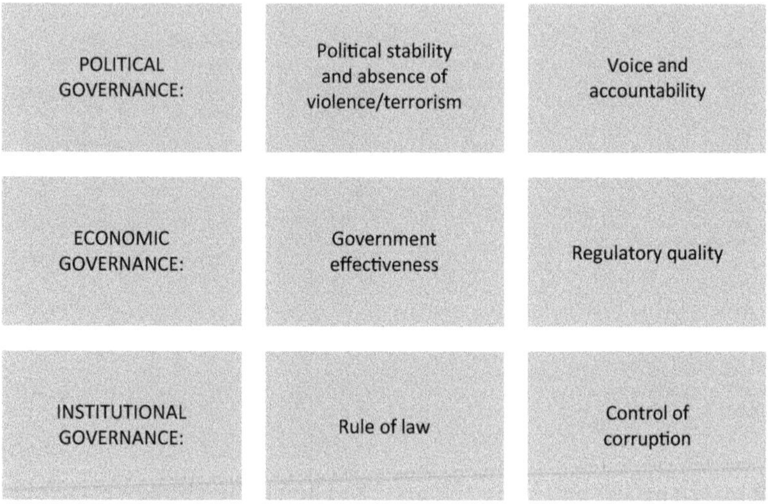

Figure 5.2 Worldwide Governance Indicators (WGI)

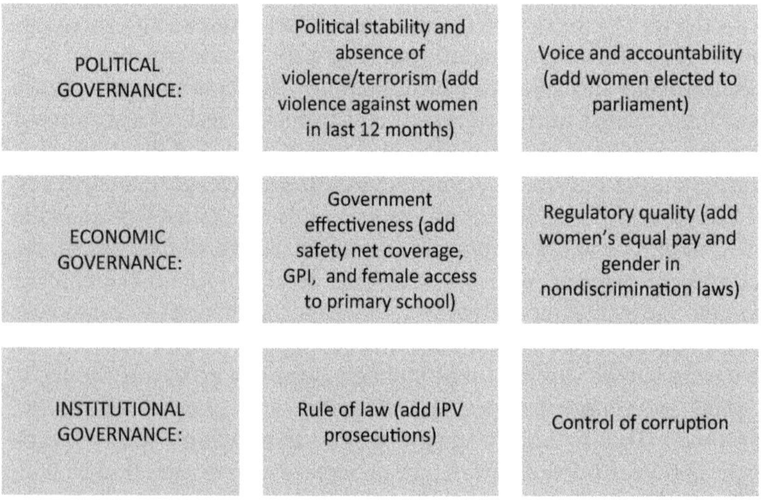

Figure 5.3 Proposed governance indicators that consider women

indexes for primary and secondary school attendance, female access to primary school, the existence of legislation promoting women's equal pay, gender in nondiscrimination legislation, and domestic violence legislation. These variables are examined at three distinct time points. All are measured at 2010–2011; WGI governance factors are measured in 2010 and 2014; and outcome factors were measured at both 2010–2011 and 2015–2016 to examine whether countries (and time periods) with better governance mean better lives for women.[1]

Describing the data

World Bank WGI and annual estimates are shown in Table 5.1.[2] According to the World Bank, estimates of governance performance range from approximately -2.5 (weak) to 2.5 (strong). In terms of political governance, as shown in Table 5.1, for all three countries, voice and accountability is weak. And it is the weakest (close to -2.5) in Zimbabwe in 2010–2011, though it is weak in 2014 as well. In comparison to the rest of the continent, only Somalia, Eritrea, and Libya have such low voice and accountability estimates in 2010.[3] And ten African nations had lower voice and accountability estimates in 2014. Estimates for political stability and absence of violence and/or terrorism were similar in that they are low (weak performance) for all three countries, and with Zimbabwe showing the lowest figures in 2010 and 2014, though estimates for 2014 are similar between Tanzania (-.60) and Zimbabwe (-.71).

Moving on to economic governance, government effectiveness is weak in all three countries. But it is weakest in Zimbabwe, with little difference between 2010 (-1.51) and 2014 (-1.21). Yet, while it improved slightly in Zimbabwe from 2010 to 2014, it became worse in Malawi (by 22 points) and Tanzania (by six points) in the same time period. Regulatory quality was also lowest in Zimbabwe, scraping the bottom score of -2.07 in 2010 and improving slightly to -1.9 in 2014. Only Somalia was worse in 2010, while several countries had similar or worse estimates of regulatory quality in 2014.

As for institutional governance, Zimbabwe had very low scores for the rule of law (-1.82 in 2010 and -1.43 in 2014) compared to Malawi and Tanzania. And control of corruption estimates in Zimbabwe were -1.37 in 2010 and -1.38 in 2014, compared to Malawi's 2010 estimate of -.49 and Tanzania's estimate of -.54 in the same year, and a more than 20-point drop in both estimates in 2014.

Table 5.2 provides descriptive statistics on all variables (independent and dependent) for the five models listed in Figure 5.1, including CPIAs (CPIA variables developed by the African Development Bank Group [AfDB]). A few are described below.

Table 5.1 Worldwide Governance Indicators of political, economic, and institutional governance: Malawi, Tanzania, and Zimbabwe (2010 and 2014)

Country	*Political governance*				*Economic governance*				*Institutional governance*			
	Voice and accountability		*Political stability and absence of violence/terrorism*		*Government effectiveness*		*Regulatory quality*		*Rule of law*		*Control of corruption*	
	2010 estimate	*2014 estimate*	*2010 estimate*	*2014 estimate*	*2010 estimate*	*2014 estimate*	*2010 estimate*	*2014 estimate*	*2010 estimate*	*2014 estimate*	*2010 estimate*	*2014 estimate*
Malawi	–0.19	–0.06	0.04	0.08	–0.41	–0.63	–0.57	–0.80	–0.13	–0.29	–0.49	–0.75
Tanzania	–0.13	–0.19	0.01	–0.60	–0.60	–0.66	–0.42	–0.33	–0.49	–0.40	–0.54	–0.75
Zimbabwe	–1.48	–1.26	–1.10	–0.71	–1.51	–1.21	–2.07	–1.90	–1.82	–1.43	–1.37	–1.39

Source: http://databank.worldbank.org/data

Note: According to the World Bank, estimates of governance range from approximately -2.5 (weak) to 2.5 (strong) governance performance.

Table 5.2 Descriptive statistics for all variables in IDHS and WDI joint analysis file

IDHS or WDI variable	*All three countries, two time periods each*			
	N	*Minimum*	*Maximum*	*Mean or proportion*
ICT				
Household has landline telephone	90,113	0	8	0.03
Household has cell phone	56,954	0	1	0.64
Household has personal computer	90,113	0	98	0.06
Female respondent owns cell phone*	47,783	0	1	0.47
Respondent uses own cell phone for financial transactions*	47,783	0	9	0.22
Respondent used the internet in the past year*	47,783	0	98	0.11
Respondent-frequency of using the internet in the past month*	47,783	0	3	0.07
Mobile cellular subscriptions (total number nationwide)	90,113	3,117,364	39,665,600	12,979,582
Mobile-cellular subscriptions (per 100 people)	90,113	20.76	84.79	47.42
Knowledge diffusion				
Literacy—reads passage easily	90,113	0	3	0.69
Female respondent, highest educational level, secondary or further	90,113	0	1	0.32
Woman's highest educational level	90,113	0	98	6.62
Primary school completion rate, female (percentage of relevant age group)	90,113	68.49	90.93	78.66
Pupil-teacher ratio, primary education	90,113	35.86	79.27	58.92
School enrollment, primary and secondary (gross), gender parity index (GPI)	90,113	0.9	1	0.94
Government effectiveness (economic governance)				
Coverage of social safety net programs (percentage of population)	90,113	16.11	41.74	25.4

(*continued*)

Table 5.2 (cont.)

IDHS or WDI variable	*All three countries, two time periods each*			
	N	*Minimum*	*Maximum*	*Mean or proportion*
CPIA equity of public resource use rating (1=low to 6=high)	90,113	2	4	3.42
CPIA quality of public administration rating (1=low to 6=high)	90,113	2	3.5	2.91
Law mandates equal remuneration for females and males for work of equal value (1=yes; 0=no)	90,113	0	1	0.53
Nondiscrimination clause mentions gender in the constitution (1=yes; 0=no)	90,113	0	1	0.26
Political governance				
CPIA transparency, accountability, and corruption in the public sector rating (1=low to 6=high)	90,113	1.5	3	2.54
Proportion of seats held by women in national parliaments (percentage)	90,113	15	36.6	23.71
Institutional governance				
CPIA gender equality rating (1=low to 6=high)	90,113	2.5	4	3.45
CPIA policies for social inclusion/equity cluster average (1=low to 6=high)	90,113	1.8	3.7	3.37
CPIA property rights and rule-based governance rating (1=low to 6=high)	90,113	1.5	3.5	3.13
Legislation exists on domestic violence (1=yes; 0=no)	90,113	0	1	0.38
Control variables				
Percentage of women aged 15–49 subjected to physical and/or sexual violence in the last 12 months	90,113	21.2	36.8	26.86
Urban residence	90,113	1	2	0.26
Age of female respondent	90,113	15	49	28.3
Resident is a visitor (versus usual resident)	90,113	1	8	0.03
Female respondent is household head	90,113	1	3	0.17
Proportion of heads of household who are male	90,113	1	2	0.69

IDHS or WDI variable	All three countries, two time periods each			
	N	Minimum	Maximum	Mean or proportion
Relationship structure in household, 2 adults of opposite sex	90,113	0	5	0.34
Relationship structure in household, 3+ related adults	90,113	0	5	0.44
Woman is currently married	90,113	0	8	0.57
Respondent agrees wife beating is justified if woman refuses to have sex	90,113	0	8	0.14
Barrier to woman's health care: lack of money for treatment a big problem	90,113	10	98	0.46
Barrier to woman's health care: distance to facility a big problem	90,113	10	98	0.44
Individuals using the internet	90,113	2.26	16.36	6.69

* Data collected in last DHS only (Malawi 2016, Tanzania 2015, and Zimbabwe 2015)

Table 5.2a Malawi (2010 and 2016)

IDHS or WDI variable	Malawi (2010)		Malawi (2016)	
	N	Mean or proportion	N	Mean or proportion
ICT				
Household has landline telephone	23,020	0.026	24,562	0.026
Household has cell phone	23,020	0.44	24,562	0.63
Household has personal computer	NA	NA	24,562	0.04
Female respondent owns cell phone*	NA	NA	24,562	0.349
Respondent uses own cell phone for financial transactions*	NA	NA	24,562	0.091
Respondent used the internet in the past year*	NA	NA	24,562	0.056
Respondent used the internet nearly daily in the past month*	NA	NA	24,562	0.031
Mobile-cellular subscriptions (total number nationwide)	23,020	3,117,364	24,562	6,566,707

(continued)

Table 5.2a (cont.)

IDHS or WDI variable	*Malawi (2010)*		*Malawi (2016)*	
	N	*Mean or proportion*	*N*	*Mean or proportion*
Mobile-cellular subscriptions (per 100 people)	23,020	20.76	24,562	37.94
Knowledge diffusion				
Literacy—reads passage easily	23,020	0.585	24,562	0.649
Female respondent, highest educational level, secondary or further	23,020	0.172	24,562	0.247
Female respondent, total years education	23,020	5.35	24,562	6.14
Primary school completion rate, female (percentage of relevant age group)	23,020	68.49	24,562	80.11
Pupil-teacher ratio, primary education	23,020	79.27	24,562	69.51
School enrollment, primary and secondary (gross), gender parity index (GPI)	23,020	0.9	24,562	0.9
Economic governance				
Coverage of social safety net programs (percentage of population)	23,020	19.8	24,562	41.74
CPIA equity of public resource use rating (1=low to 6=high)	23,020	3.5	24,562	3.5
CPIA quality of public administration rating (1=low to 6=high)	23,020	3.5	24,562	2.5
Law mandates equal remuneration for females and males for work of equal value (1=yes; 0=no)	23,020	0	24,562	1
Nondiscrimination clause mentions gender in the constitution (1=yes; 0=no)	23,020	0	24,562	0
Political governance				
CPIA transparency, accountability, and corruption in the public sector rating (1=low to 6=high)	23,020	3	24,562	2.5
Proportion of seats held by women in national parliaments (percentage)	23,020	20.8	24,562	16.7

IDHS or WDI variable	*Malawi (2010)*		*Malawi (2016)*	
	N	*Mean or proportion*	*N*	*Mean or proportion*
Institutional governance				
CPIA gender equality rating (1=low to 6=high)	23,020	3.5	24,562	3.5
CPIA policies for social inclusion/equity cluster average (1=low to 6=high)	23,020	3.5	24,562	3.5
CPIA property rights and rule-based governance rating (1=low to 6=high)	23,020	3.5	24,562	3.5
Legislation exists on domestic violence (1=yes; 0=no)	23,020	0	24,562	1
Control variables				
Percentage of women aged 15–49 subjected to physical and/or sexual violence in the past 12 months	23,020	21.2	24,562	21.2
Urban residence	23,020	0.13	24,562	0.21
Age of female respondent	23,020	28.12	24,562	28.14
Resident is a visitor (versus usual resident)	23,020	0.02	24,562	0.01
Female respondent is household head	23,020	0.16	24,562	0.18
Proportion of heads of household who are male	23,020	0.73	24,562	0.70
Relationship structure in household, 2 adults of opposite sex	23,020	0.39	24,562	0.37
Relationship structure in household, 3+ related adults	23,020	0.42	24,562	0.43
Woman is currently married	23,020	0.59	24,562	0.61
Respondent agrees wife beating is justified if woman refuses to have sex	23,020	0.06	24,562	0.08
Barrier to woman's health care: lack of money for treatment a big problem	23,020	0.53	24,562	0.48
Barrier to woman's health care: distance to facility a big problem	23,020	0.58	24,562	0.51
Individuals using the internet	23,020	2.26	24,562	9.3

* Data collected in last DHS only (Malawi 2016, Tanzania 2015, and Zimbabwe 2015)

Table 5.2b Tanzania (2010 and 2016)

IDHS or WDI variable	*Tanzania (2010)*		*Tanzania (2015)*	
	N	*Mean or proportion*	*N*	*Mean or proportion*
ICT				
Household has landline telephone	10,139	0.02	13,266	0.01
Household has cell phone	10,139	0.57	13,266	0.86
Household has personal computer	NA	NA	13,266	0.05
Female respondent owns cell phone*	NA	NA	13,266	0.52
Respondent uses own cell phone for financial transactions*	NA	NA	13,266	0.32
Respondent used the internet in past year*	NA	NA	13,266	0.02
Respondent used internet nearly daily in last month*	NA	NA	13,266	0.03
Mobile-cellular subscriptions (total number nationwide)	10,139	20,983,853	13,266	39,665,600
Mobile-cellular subscriptions (per 100 people)	10,139	46.66	13,266	75.86
Knowledge diffusion				
Literacy—reads passage easily	10,139	0.66	13,266	0.72
Female respondent, highest educational level, secondary or further	10,139	0.23	13,266	0.26
Female respondent, total years education	10,139	5.93	13,266	6.44
Primary school completion rate, female (percentage of relevant age group)	10,139	77.1	13,266	77.1
Pupil-teacher ratio, primary education	10,139	50.76	13,266	43.06
School enrollment, primary and secondary (gross), gender parity index (GPI)	10,139	0.96	13,266	1
Economic governance				
Coverage of social safety net programs (percentage of population)	10,139	16.11	13,266	19.78
CPIA equity of public resource use rating (1=low to 6=high)	10,139	4	13,266	4

IDHS or WDI variable	*Tanzania (2010)*		*Tanzania (2015)*	
	N	*Mean or proportion*	*N*	*Mean or proportion*
CPIA quality of public administration rating (1=low to 6=high)	10,139	3	13,266	3.5
Law mandates equal remuneration for females and males for work of equal value (1=yes; 0=no)	10,139	0	13,266	1
Nondiscrimination clause mentions gender in the constitution (1=yes; 0=no)	10,139	0	13,266	1
Political governance				
CPIA transparency, accountability, and corruption in the public sector rating (1=low to 6=high)	10,139	2.5	13,266	3
Proportion of seats held by women in national parliaments (percentage)	10,139	30.7	13,266	36.6
Institutional governance				
CPIA gender equality rating (1=low to 6=high)	10,139	3.5	13,266	3.5
CPIA policies for social inclusion/equity cluster average (1=low to 6=high)	10,139	3.7	13,266	3.7
CPIA property rights and rule-based governance rating (1=low to 6=high)	10,139	3.5	13,266	3.5
Legislation exists on domestic violence (1=yes; 0=no)	10,139	0	13,266	0
Control variables				
Percentage of women aged 15–49 subjected to physical and/or sexual violence in the past 12 months	10,139	36.8	13,266	36.8
Urban residence	10,139	0.256	13,266	0.31
Age of female respondent	10,139	28.72	13,266	28.69
Resident is a visitor (versus usual resident)	10,139	0.05	13,266	0.06
Female respondent is household head	10,139	0.11	13,266	0.11
Proportion of heads of household who are male	10,139	0.78	13,266	0.77

(*continued*)

Table 5.2b (cont.)

IDHS or WDI variable	*Tanzania (2010)*		*Tanzania (2015)*	
	N	*Mean or proportion*	*N*	*Mean or proportion*
Relationship structure in household, 2 adults of opposite sex	10,139	0.31	13,266	0.31
Relationship structure in household, 3+ related adults	10,139	0.51	13,266	0.45
Woman is currently married	10,139	0.58	13,266	0.46
Respondent agrees wife beating is justified if woman refuses to have sex	10,139	0.27	13,266	0.3
Barrier to woman's health care: lack of money for treatment a big problem	10,139	0.23	13,266	0.49
Barrier to woman's health care: distance to facility a big problem	10,139	0.18	13,266	0.40
Individuals using the internet	10,139	2.9	13,266	5.36

* Data collected in last DHS only (Malawi 2016, Tanzania 2015, and Zimbabwe 2015)

Table 5.2c Zimbabwe (2010–2011 and 2015–2016)

IDHS or WDI variable	*Zimbabwe (2010–2011)*		*Zimbabwe (2015–2016)*	
	N	*Mean or proportion*	*N*	*Mean or proportion*
ICT				
Household has landline telephone	9,171	0.049	9,955	0.5
Household has cell phone	9,171	0.666	9,955	0.92
Household has personal computer	9,171	0.044	9,955	0.15
Female respondent owns cell phone*	NA	NA	9,955	0.72
Respondent uses own cell phone for financial transactions*	NA	NA	9,955	0.40
Respondent used the internet in the past year*	NA	NA	9,955	0.28

IDHS or WDI variable	*Zimbabwe (2010–2011)*		*Zimbabwe (2015–2016)*	
	N	*Mean or proportion*	*N*	*Mean or proportion*
Respondent used the internet nearly daily in the past month*	NA	NA	9,955	0.20
Mobile-cellular subscriptions (total number nationwide)	9,171	7,700,000	9,955	12,757,410
Mobile-cellular subscriptions (per 100 people)	9,171	58.88	9,955	84.79
Knowledge diffusion				
Literacy—reads passage easily	9,171	0.87	9,955	0.87
Female respondent, highest educational level, secondary or further	9,171	0.64	9,955	0.67
Female respondent, total years education	9,171	8.95	9,955	9.52
Primary school completion rate, female (percentage of relevant age group)	9,171	90.93	9,955	90.93
Pupil-teacher ratio, primary education	9,171	35.86	9,955	36.41
School enrollment, primary and secondary (gross), gender parity index (GPI)	9,171	0.99	9,955	0.98
Economic governance				
Coverage of social safety net programs (percentage of population)	9171	19.98	9,955	19.98
CPIA equity of public resource use rating (1=low to 6=high)	9,171	2	9,955	3
CPIA quality of public administration rating (1=low to 6=high)	9,171	2	9,955	2.5
Law mandates equal remuneration for females and males for work of equal value (1=yes; 0=no)	9,171	0	9,955	1
Nondiscrimination clause mentions gender in the constitution (1=yes; 0=no)	9,171	0	9,955	1

(*continued*)

Table 5.2c (cont.)

IDHS or WDI variable	*Zimbabwe (2010–2011)*		*Zimbabwe (2015–2016)*	
	N	*Mean or proportion*	*N*	*Mean or proportion*
Political governance				
CPIA transparency, accountability, and corruption in the public sector rating (1=low to 6=high)	9,171	1.5	9,955	2
Proportion of seats held by women in national parliaments (percentage)	9,171	15	9,955	31.5
Institutional governance				
CPIA gender equality rating (1=low to 6=high)	9,171	2.5	9,955	4
CPIA policies for social inclusion/equity cluster average (1=low to 6=high)	9,171	1.8	9,955	3.4
CPIA property rights and rule-based governance rating (1=low to 6=high)	9,171	1.5	9,955	2
Legislation exists on domestic violence (1=yes; 0=no)	9,171	0	9,955	1
Control variables				
Percentage of women aged 15–49 subjected to physical and/or sexual violence in the last 12 months	9,171	30.5	9,955	27.2
Urban residence	9,171	0.38	9,955	0.45
Age of female respondent	9,171	28.15	9,955	28.52
Resident is a visitor (versus usual resident)	9,171	0.04	9,955	0.05
Female respondent is household head	9,171	0.28	9,955	0.25
Proportion of heads of household who are male	9,171	0.50	9,955	0.56
Relationship structure in household, 2 adults of opposite sex	9,171	0.27	9,955	0.28
Relationship structure in household, 3+ related adults	9,171	0.47	9,955	0.41
Woman is currently married	9,171	0.58	9,955	0.57

IDHS or WDI variable	*Zimbabwe (2010–2011)*		*Zimbabwe (2015–2016)*	
	N	*Mean or proportion*	*N*	*Mean or proportion*
Female respondent agrees wife beating is justified if woman refuses to have sex	9,171	0.16	9,955	0.13
Barrier to woman's health care: lack of money for treatment a big problem	9,171	0.51	9,955	0.41
Barrier to woman's health care: distance to facility a big problem	9,171	0.35	9,955	0.30
Individuals using the internet	9,171	6.4	9,955	16.36

* Data collected in last DHS only (Malawi 2016, Tanzania 2015, and Zimbabwe 2015)

CPIA variables

The percentage of the population covered by social safety net programs in Malawi, Tanzania, and Zimbabwe ranged from 16% to 42%. Equity in public resource use ranged from 4 in Tanzania to 2 in Zimbabwe (on a six-point scale, with 6=high). Quality of public administration ranged from 3.5 in Tanzania to 2 in Zimbabwe (where 1=low and 6=high). And the CPIA gender equality rating ranged from 4 (Zimbabwe 2015–2016) to 2.5 (Zimbabwe 2010). Malawi tended to lie in the middle range of these CPIA variables.

ICT variables from the DHS in Malawi, Tanzania, and Zimbabwe (2010–2016)

Few households had landlines (3%); however, 64% of households had cell phones. And 47% of female respondents (aged 15–49) had their own cell phones. Almost half (22%) used cell phones for financial transactions. A total of 11% of women had used the internet in the past year.

Knowledge diffusion from the DHS

In 2010–2016 in Malawi, Tanzania, and Zimbabwe, most women had completed seven years of education, and about one-third had only completed secondary-level education. However, the primary school completion rate, meaning that the percentage of the relevant age group was 79%. The gender parity index in primary and secondary school was .94 (an index of value less than 1 indicates that indicator values are higher for males than for females).

Outcome variables

Most women were able to read a passage easily (69%). A little under half of women felt that distance or money were barriers to accessing women's health care. And 14% agreed that "wife beating is justified" if the woman refuses to have sex. This ranged from 30% in Tanzania (2015) to 6% in Malawi (2016).

Control variables

Only 26% of respondents lived in urban areas. As few as 17% of households were headed by the female respondent. A little over one-third (34%) of households appeared to be composed of a heteronormative nuclear family and 57% of women were married.

Factors related to women's ownership of cell phones[4]

Table 5.3 shows a statistically significant model that includes urban residence, country, family structure, marriage, and age of the respondent as control factors that are all related to female respondents' cell phone ownership in 2015–2016 (the variable is not available in the 2010–2011 datasets). As shown, women in urban areas are more than two times more likely to own cell phones. And women in Malawi and Tanzania are less likely than women in Zimbabwe to own cell phones. And if the woman respondent is the head of her household, she is almost two times more likely to own a cell phone compared to women who were not household heads. Further, if a cell phone is owned by a member of the household, the woman is 25 times more likely to own a phone herself, net of other determinants. And educational levels are also related to women's cell phone ownership. For every additional year of education, female respondents were three times more likely to own a cell phone.

What are indicators of good governance for women?

A number of NGOs and women's grassroots organizations promote the importance of effective governance that benefits women. Some

Table 5.3 Analysis of determinants of women respondents' cell phone ownership

		Exp(B)	*Sig.*
Control variables	**MODEL 1**		
Urban residence		2.333	***
Country: Malawi		0.592	***
Country: Tanzania		0.892	**
Country: Zimbabwe (omitted category)			
Resident is a visitor (versus usual resident)		0.68	***
Female respondent is household head		1.822	***
Proportion of heads of household who are male		0.718	***
Relationship structure in household, 2 adults of opposite sex		0.805	***
Relationship structure in household, 3+ related adults		0.623	***
Woman is currently married		1.355	***
Age of female respondent		1.069	***
ICT	**MODEL 2**		
Household has cell phone		24.951	***
Knowledge diffusion	**MODEL 3**		
Female respondent, did not receive formal education		2.512	***
Female respondent, primary educational level (omitted category)			
Female respondent, secondary or furthereducational level		2.906	***
Constant		0.005	***

Notes:

B = coefficient, Sig. = significance level, Exp(B) = odds ratio.

Significance levels are grouped into two categories: $^{**}p<0.01$; $^{***}p<0.001$.

Source: Zerai's analysis of data extract performed by DHS implementing partners and ICF International. DHS: Malawi 2010 and 2015–2016; Tanzania 2010 and 2015–2016; Zimbabwe 2010 and 2015–2016. Data extract from DHS Recode files. IPUMS-Demographic and Health Surveys (IPUMS-DHS), version 3.0, Minnesota Population Center and ICF International (Distributors). Accessed at: http://idhsdata.org

include Advancing Girls Education in Africa, Akina Mama Wa Afrika, BAOBAB for Women's Human Rights, Girls Power Initiative, Tanzania Media Women's Association, Women's Fund Tanzania, and WOZA. On the basis of both scholarly and activist literature, I identified eight variables that might be indicators of good governance (available in various types of international governance data) for women to examine in this chapter. These are as follows:

- Violence against women in last the 12 months (Raaber 2007; Jewkes 2010; Viriri et al. 2011; Mtambalike 2013; Bazargan-Hejazi et al. 2013; Daley 2014; Wekwete et al. 2014; Gwede 2017; Henderson et al. 2017; Jakobsen 2016).
- Women's political participation and seats in parliament (Gaidzanwa 1999a; Semu 2002, 2005; Campbell 2003; WOZA 2007; Women in Politics Support Unit 2008; Gama 2013; Chingwete et al. 2014; Masina 2017; Mutungi 2016; Kayuni and Chikadza 2016; Inge Amundsen and Kayuni 2016).
- Safety net coverage (Gaidzanwa 1998; Zerai 2014).
- Primary and secondary education gender parity index, and female literacy and access to primary school (Kisanga and Katunzi 1997; Gaidzanwa 1997, Mlyakado 2013; Sanya and Odero 2017).
- Women's equal pay (Gaidzanwa 1992, 1996; Wekwete 2014).
- Gender in nondiscrimination laws, including land rights (Gaidzanwa 1994; Semu 2005, 2011; Gaidzanwa 2011; Madebwe and Madebwe 2011; Campbell 2003; Mafa 2015; Rutashobya 1998; Mbilinyi 1991).
- IPV prosecutions (Usdin et al. 2000; Seedat et al. 2009; Viriri et al. 2011).
- Access to women's health and health care (WOZA 2007; Gaidzanwa 2008; Ntata et al. 2008; Wekwete and Manyeruke 2012; Lund et al. 2012; Oyeyemi and Wynn 2014).

Consistent with Kaufmann et al's 2010 conceptualization of WGI, I grouped these into political, economic, and institutional governance categories. I describe the data for Malawi, Tanzania, and Zimbabwe (2010–2011 and 2015–2016) below.

Political governance

For political stability and absence of violence/terrorism, I examined the percentage of women aged 15–49 subjected to physical and/or sexual violence in the last 12 months. This ranged from 21% to 37%, a range that is high. For voice and accountability, women's political activity is an appropriate concept to examine. While organizational affiliation is available in the DHS, it is not available in the three countries under examination, and nor is voting behavior. However, the proportion of seats held by women in national parliaments is available from the World Bank and ranges from 15 to 37.

Economic governance

Examining regulatory quality, women's equal pay, and gender in nondiscrimination laws are appropriate variables. For Malawi, Tanzania, and Zimbabwe, none had laws mandating equal remuneration for females and males for work of equal value in 2010–2011, whereas they all did by 2015–2016. And while the nondiscrimination clause does not mention gender in the Constitution in Malawi for either time period, it is mentioned in Tanzania and Zimbabwe during the latter time period (2015–2016). Moving to government effectiveness, 16–42% of the populations in each country enjoy social safety net coverage.

Institutional governance

For control of corruption, I do not have direct measures that focus on women in CPIA, WGI or DHS; but for the rule of law, I examine whether legislation exists on domestic violence. This legislation did not exist in 2010–2011 for Malawi, Tanzania, or Zimbabwe, and it only existed in the latter period (2015–2016) for Malawi and Zimbabwe.

I examined the extent to which ICT and knowledge diffusion are related to each of these indicators of good governance in order to explore whether the WGI may be improved by considering governance factors that are relevant to women's lives.

What is the difference between WGI and CPIA?

While the WGI from the World Bank (Kaufmann et al. 2010) capture six key dimensions of governance (voice and accountability, political stability and lack of violence, government effectiveness, regulatory quality, rule of law, and control of corruption) for 200 countries[5] between 1996 and the present day, the CPIA of the World Bank and adopted in a modified form by the AfDB:

> is a rating system designed to assess the performance of countries' policy and institutional frameworks in terms of their capacity to ensure the efficient utilization of scarce resources for achieving sustainable and inclusive growth. It is based on the scoring of 18 criteria covering different aspects of development such as: economic and public-sector management; structural policies; social inclusion and equity.
>
> (African Development World Bank 2015, p. 3)[6]

For WGI, estimates range from -2.5 to +2.5, as noted before, whereas for CPIA, the scores range between 1 (very weak) and 6 (very strong); both are updated every year. Therefore, time series data is available for both sets of measures.

One substantive difference in the measures from the standpoint of conducting gender-sensitive analyses is that the WGI do not include strong measures that capture women's experiences. I argue that there is considerable room to capture these, especially in relation to political stability, voice and accountability, regulatory quality, government effectiveness, and the rule of law, but even corruption control could be addressed by examining sexual harassment policies and the extent to which they are applied, if they are present at all. Fortunately, a global variable in the CPIA dataset captures gender equity, and specific measures associated with gender that correspond to five of the six indices in the WGI are also available in the CPIA.

Below, I analyze these both individually and as an aggregate gender equality variable.[7] For the CPIA gender equality variable, considerations are given to circumstances such as: whether CEDAW has been ratified; whether significant differences exist in female to male enrollment in primary or secondary education; whether gender disparities exist in participation and remuneration in the labor force, business ownership, land tenure, property ownership, and inheritance practices; recent efforts to make laws or policies more supportive of gender equality in education; access to antenatal and delivery care, and access to family planning services; whether violence against women is considered a crime; and, where laws are expected for any of the above, the extent to which they are enforced. However, other characteristics of gender equality such as the political participation of women and sexual harassment policy are not present as a part of the aggregate CPIA gender equality variable.

Do ICT and knowledge diffusion improve the lives of women aged 15–49 in Malawi, Tanzania, and Zimbabwe?

In order to examine the impact of good governance on women's lives, my conceptual framework explores whether ICT, knowledge diffusion, and good governance together affect various outcomes for women (see Figure 5.1). So, the first three models ask whether ICT (cell phones) and knowledge diffusion improve women's lives. To test this, I examine their impact on three constructs: women's literacy,[8] women's health care access (which is measured as whether money or distance are barriers to health care), and beliefs about intimate partner violence. The results are shown in Table 5. 4. Women's outcomes in Malawi, Tanzania, and Zimbabwe are examined together in 2010–2011 and 2015–2016.

Table 5.4a Analysis of models 1–5: control factors, ICT, knowledge diffusion, gender-inclusive governance factors, and aggregate WDI as determinants of women's outcomes in Malawi, Tanzania, and Zimbabwe (2010–2011 and 2015–2016)

Dependent variable: cannot read passage at all	*Control variables and cannot read passage*		*ICT and cannot read passage at all*		*Knowledge diffusion and cannot read passage*		*Inclusive governance and cannot read passage*		*WDI governance and cannot read passage*	
	Model 1		*Model 2*		*Model 3*		*Model 4*		*Model 5*	
	Exp(B)	*Sig.*	*Exp(B)*	*Sig.*	*Exp(B)*	*Sig.*	*Exp(B)*	*Sig.*	*Exp(B)*	*Sig.*
Control variables										
Urban residence	0.305	***	0.407	***	0.7	***	0.703	***	0.703	***
Survey year	1.217	***	0.967		0.902	***	0.739	***	0.725	***
Country: Malawi	5.595	***	4.704	***	1.747	***	1.151	***	0.808	
Country: Tanzania	5.1	***	4.955	***	1.119	*	0.785	***	0.567	**
Country: Zimbabwe (omitted category)										
Resident is a visitor (versus usual resident)	1.192	**	1.092		0.988		1.001	***	1.001	
Female respondent is household head	1.318	***	1.267	***	1.085		1.092	***	1.092	
Proportion of heads of household who are male	1.195	***	1.311	***	1.169	***	1.162	***	1.162	***
Relationship structure in household, 2 adults of opposite sex	1.239	***	1.151	***	1.064		1.065	***	1.066	

(*continued*)

Table 5.4a (cont.)

Dependent variable: cannot read passage at all	*Control variables and cannot read passage*		*ICT and cannot read passage at all*		*Knowledge diffusion and cannot read passage*		*Inclusive governance and cannot read passage*		*WDI governance and cannot read passage*	
	Model 1		*Model 2*		*Model 3*		*Model 4*		*Model 5*	
	Exp(B)	*Sig.*	*Exp(B)*	*Sig.*	*Exp(B)*	*Sig.*	*Exp(B)*	*Sig.*	*Exp(B)*	*Sig.*
Relationship structure in household, 3+ related adults	0.969		1.002		0.925	*	0.933	***	0.934	*
Woman is currently married	1.115	***	1.176	***	1.041		1.058	***	1.056	*
Age of female respondent	1.046	***	1.047	***	1.011	***	1.011	***	1.011	***
ICT										
Household has cell phone			0.375	***	0.557	***	0.549	***	0.549	***
Knowledge diffusion										
Female respondent, did not receive formal education					0.017	***	0.017	***	0.017	***
Female respondent, primary educational level (omitted category)										
Female respondent, secondary or further education					0.029	***	0.029	***	0.029	***
Evidence of gender-inclusive-governance										

Primary school completion rate, female (percentage of relevant age group)							0.976	***	0.978	***
Proportion of seats held by women in national parliaments (percentage)										
Governance factor from WGI										
Includes voice and accountability and political stability and absence of violence/terrorism (Pol Gov); government effectiveness and regulation quality (Econ Gov); and rule of law and control of corruption (Inst'l Gov)									1.175	
Constant	0.013	***	0.026	***	10.418	***	108.021	***	114.985	***

Notes:
B = coefficient; Sig. = significance level; Exp(B) = odds ratio.

Significance levels are grouped into three categories: *$p<0.05$; **$p<0.01$; ***$p<0.001$.

Table 5.4b Analysis of models 1–5: determinants of women's literacy

Dependent variable: reads passage easily	*Control variables and reads passage easily*		*ICT and reads passage easily*		*Knowledge diffusion and reads passage easily*		*Inclusive governance and reads passage easily*		*WDI governance and reads passage easily*	
	Model 1		*Model 2*		*Model 3*		*Model 4*		*Model 5*	
	Exp(B)	*Sig.*	*Exp(B)*	*Sig.*	*Exp(B)*	*Sig.*	*Exp(B)*	*Sig.*	*Exp(B)*	*Sig.*
Control variables										
Urban residence	3.085	***	2.374	***	1.36	***	1.355	***	1.354	***
Survey year	0.85	***	1.052	**	1.123	***	1.323	***	1.335	***
Country: Malawi	0.284	***	0.334	***	0.975		1.394	***	2.992	***
Country: Tanzania	0.367	***	0.381	***	1.793	***	2.431	*	4.922	***
Country: Zimbabwe (omitted category)										
Resident is a visitor (versus usual resident)	0.802	***	0.866	**	0.957		0.946	***	0.945	
Female respondent is household head	0.843	***	0.874	***	1.012		1.006	***	1.005	
Proportion of heads of household who are male	0.893	***	0.822	***	0.922	*	0.927	***	0.928	*
Relationship structure in household, 2 adults of opposite sex	0.805	***	0.86	***	0.938	*	0.937	***	0.935	*
Relationship structure in household, 3+ related adults	1.018		0.986		1.056		1.048	***	1.048	
Woman is currently married	0.915	***	0.872	***	0.978	***	0.965	***	0.97	
Age of female respondent	0.959	***	0.959	***	0.987	***	0.987	***	0.987	***
ICT										
Household has cell phone			2.492	***	1.619	***	1.637	***	1.638	***

Knowledge diffusion										
Female respondent, did not receive formal education					65.658	***	65.911	***	65.671	***
Female respondent, primary educational level (omitted category)										
Female respondent, secondary or further education					19.522	***	19.53	***	19.719	***
Evidence of gender-inclusive governance										
Primary school completion rate, female (percentage of relevant age group)							1.022	***	1.012	**
Proportion of seats held by women in national parliaments (percentage)										
Governance factor from WGI										
Includes voice and accountability and political stability and absence of violence/terrorism (Political Governance); government effectiveness and regulation quality (Economic Governance); and rule of law and control of corruption (Institutional Governance)									0.691	***
Constant	27.207	***	13.771	***	0.029	***	0.004	***	0.004	***

Table 5.4c Analysis of models 1–5: determinants of women's opinions about wife beating

Dependent variable: female respondent agrees wife beating is justified if refuses sex	*Control variables and agrees wife beating is justified*		*ICT and respondent agrees wife beating is justified*		*Knowledge diffusion and respondent agrees wife beating is justified*		*Inclusive governance and respondent agrees wife beating is justified*		*WDI governance and respondent agrees wife beating is justified*	
	Model 1		*Model 2*		*Model 3*		*Model 4*		*Model 5*	
	Exp(B)	*Sig.*	*Exp(B)*	*Sig.*	*Exp(B)*	*Sig.*	*Exp(B)*	*Sig.*	*Exp(B)*	*Sig.*
Control variables										
Urban residence	0.508	***	0.557	***	0.64	***	0.639	***	0.64	***
Survey year	0.88	***	0.803	***	0.806	***	0.709	***	0.797	***
Country: Malawi	0.371	***	0.342	***	0.256	***	0.225	***	.584	*
Country: Tanzania	2.179	***	2.138	***	1.576	***	2.039	***	3.947	***
Country: Zimbabwe (omitted category)										
Resident is a visitor (versus usual resident)	1.235	***	1.21	***	1.159	**	1.149	**	1.151	**
Female respondent is household head	1.037		01.019		.994		0.987		0.988	
Proportion of heads of household who are male	1.05		1.076	*	1.054		1.063		1.062	
Relationship structure in household, 2 adults of opposite sex	1.046		1.015		0.973		0.968		0.966	
Relationship structure in household, 3+ related adults	1.021		1.029		1.018		1.006		1.009	
Woman is currently married	0.958		0.971		0.94	**	0.931	**	0.939	**
Age of female respondent	0.992	***	0.991	***	0.986	***	0.986	***	0.986	***

ICT										
Household has cell phone			0.702	***	0.801	***	0.811	***	0.807	***
Knowledge diffusion										
Female respondent, did not receive formal education					0.908	***	0.904	***	0.903	**
Female respondent, primary educational level (omitted category)										
Female respondent, secondary or further education					0.483	***	0.483	***	0.484	***
Evidence of gender-inclusive-governance										
Primary school completion rate, female (percentage of relevant age group)										
Proportion of seats held by women in national parliaments (percentage)							.975	***	.989	**
Governance factor from WGI										
Includes voice and accountability and political stability and absence of violence/terrorism (political governance); government effectiveness and regulation quality (economic governance); and rule of law and control of corruption (institutional governance)									.700	***
Constant	0.270	***	0.365	***	0.713	***	1.382	**	.477	**

Table 5.4d Analysis of models 1–5: determinants of distance barriers to women's medical care

Dependent variable: distance a barrier to female respondent's health care	*Control variables and distance a barrier to health care*		*ICT and distance a barrier to health care*		*Knowledge diffusion and distance a barrier to health care*		*Inclusive governance and distance a barrier to health care*		*WDI governance and distance a barrier to health care*	
	Model 1		*Model 2*		*Model 3*		*Model 4*		*Model 5*	
	Exp(B)	*Sig.*	*Exp(B)*	*Sig.*	*Exp(B)*	*Sig.*	*Exp(B)*	*Sig.*	*Exp(B)*	*Sig.*
Control variables										
Urban residence	0.223	***	0.246	***	0.27	***	0.271	***	0.272	***
Survey year	0.861	***	0.794	***	0.79	***	0.498	***	0.436	***
Country: Malawi	1.879	***	1.755	***	1.484	***	0.489	***	10.596	***
Country: Tanzania	0.745	***	0.729	***	0.598	***	0.231	***	3.801	***
Country: Zimbabwe (omitted category)										
Resident is a visitor (versus usual resident)	1.204	***	1.172	***	1.138	**	1.176	***	1.181	***
Female respondent is household head	1.088	**	1.073	*	1.055		1.073	*	1.071	*
Proportion of heads of household who are male	1.007		1.036		1.018		1.001		1.007	
Relationship structure in household, 2 adults of opposite sex	1.101	***	1.07	**	1.047		1.056	*	1.045	
Relationship structure in household, 3+ related adults	0.98		0.989		0.981		1.012		1.01	
Woman is currently married	0.939	***	0.952	**	0.935	***	0.969		0.991	
Age of female respondent	1.007	***	1.007	***	1.003	**	1.002	**	1.002	*
ICT										
Household has cell phone			0.708	***	0.776	***	0.748	***	0.737	***

Knowledge diffusion										
Female respondent, did not receive formal education					0.858	***	0.861	***	0.847	***
Female respondent, primary educational level (omitted category)										
Female respondent, secondary or further education					0.659	***	0.669	***	0.675	***
Evidence of gender-inclusive-governance										
Primary school completion rate, female (percentage of relevant age group)							0.934	***	0.884	***
Proportion of seats held by women in national parliaments (percentage)										
Governance factor from WGI										
Includes voice and accountability and political stability and absence of violence/terrorism (Political Governance); government effectiveness and regulation quality (Economic Governance); and rule of law and control of corruption (Institutional Governance)									0.199	***
Constant	0.591	***	0.79	***	1.317	***	773.811	***	5,857.157	***

Table 5.4e Analysis of models 1–5: determinants of financial barriers to women's medical care

Dependent variable: money a barrier to female respondent's health care	*Control variables and money a barrier*		*ICT and money a barrier to health care*		*Knowledge diffusion and money a barrier to health care*		*Inclusive governance and money a barrier to health care*		*WDI governance and money a barrier to health care*	
	Model 1		*Model 2*		*Model 3*		*Model 4*		*Model 5*	
	Exp(B)	*Sig.*	*Exp(B)*	*Sig.*	*Exp(B)*	*Sig.*	*Exp(B)*	*Sig.*	*Exp(B)*	*Sig.*
Control variables										
Urban residence	0.46	***	0.533	***	0.601	***	0.603	***	0.605	***
Survey year	0.874	***	0.776	***	0.77	***	0.547	***	0.46	***
Country: Malawi	1.022	***	0.918	***	0.739	***	0.309	***	11.075	***
Country: Tanzania	0.648	***	0.626	***	0.484	***	0.23	***	6.081	***
Country: Zimbabwe (omitted category)										
Resident is a visitor (versus usual resident)	1.222	***	1.18	***	1.142	**	1.171	***	1.174	***
Female respondent is household head	1.073	*	1.051		1.032		1.044		1.042	
Male sex of household head	0.84	***	0.872	***	0.852	***	0.839	***	0.845	***
Relationship structure in household, 2 adults of opposite sex	1.214	***	1.168	***	1.138	***	1.148	***	1.133	***
Relationship structure in household, 3+ related adults	1.011		1.028		1.021		1.047	*	1.045	*
Woman is currently married	0.74	***	0.754	***	0.737	***	0.756	***	0.775	***
Age of female respondent	1.019	***	1.019	***	1.014	***	1.013	***	1.013	***
ICT										
Household has cell phone			0.589	***	0.661	***	0.646	***	0.627	***

Knowledge diffusion										
Female respondent, did not receive formal education					0.803	***	0.805	***	0.788	***
Female respondent, primary educational level (omitted category)										
Female respondent, secondary or further education					0.594	***	0.6	***	0.605	***
Evidence of gender-inclusive governance										
Primary school completion rate, female (percentage of relevant age group)							0.948	***	0.888	***
Proportion of seats held by women in national parliaments (percentage)										
Governance factor from WGI										
Includes voice and accountability and political stability and absence of violence/terrorism (Political Governance); government effectiveness and regulation quality (Economic Governance); and rule of law and control of corruption (Institutional Governance)									0.151	***
Constant	0.719	***	1.114	***	2.17	***	318.203	***	3,822.135	***

Model 1 includes control variables, model 2 adds an ICT variable, and model 3 includes knowledge diffusion.

Women's literacy

Women's literacy has a strong and statistically significant relationship with control variables such as urban residence, time period, country, visitor status, family and household structure, marital status, and age of the respondent. Urban residence is related to women easily reading passages, as women from urban areas are three times more likely to easily read passages compared to women from rural areas. And women who are married and older are more likely to be unable to read a passage (as shown in model 1). Reading easily is also related to access to ICT (model 2) and knowledge diffusion (model 3). Later time periods (the 2015–2016 survey years compared to the 2010–2011 survey years) are related to an increased likelihood of improvements in the numbers of women easily reading passages, even when access to ICT is taken into account (model 2). And women who have cell phones in the household are 2.492 times more likely to read a passage easily, over and above the effect of other factors.

Beliefs about IPV

Some notable results in the models examining the effect of governance on beliefs about IPV are that women in Tanzania are more likely to agree that IPV is justified if a wife refuses to have sex compared to Malawi and Zimbabwe. Women in homes with cell phone are less likely to agree that IPV is justified if a woman refuses to have sex compared to those without cell phones.

Barriers to health care for women

Money and distance barriers to health care are 11.1 times more likely in Malawi compared to Zimbabwe, when control factors, ICT, and knowledge diffusion variables are taken into account. Visitors [nonresidents][9] are more likely to have money or distance barriers to receiving women's health care, when control, ICT and knowledge diffusion variables are taken into account. And older age is related to barriers to obtaining women's health. Cell phones in the house are related to lower numbers of women for whom money or distance are health care barriers. Higher educational levels are related to lower numbers of women for whom money or distance is a barrier to access to their own medical care.

What impact do World Bank definitions of good governance have on women's lives?

A global good governance variable, which is a factor of all six WGI, is added and presented in model 5. The global governance indicator does not have a significant effect on women's inability to read a passage, but it is related to women's ability to read a passage easily, though it is counterintuitive, in that women with higher relative levels of governance are less likely to be able to read a passage easily. Another counterintuitive finding is that women who live in areas where there are higher relative levels of governance are 17 times more likely to believe that "wife beating is justified" if a wife refuses to have sex. Two findings that correspond to the hypothesis are that financial and transportation barriers are related to higher relative levels of governance.

I was curious about whether I could add alternative proxies of governance that were relevant to women to a model that included the aggregate WGI good governance variable. My hypothesis was that the women-focused proxies for political, economic, and institutional governance would improve models. I added these as women's context, or "gender-inclusive governance" variables, to model 4. However, only two variables provided robust results that included both women-focused proxies and the aggregate good governance variable from the World Bank. Of the eight proxies noted above, these were only the primary school completion rate that considered the relevant age group of females and the proportion of seats held by women in parliament. Including primary school completion rate, and female school enrollment (calculated as the total percentage of the relevant age group), difficulty in reading is less likely, easy reading is more likely, the belief that IPV is justified if women refuse to have sex is less likely, and financial and distance barriers to women's health care are less likely (as shown in Table 5.4). Women's political participation, a proxy of political governance, significantly added to the model only for the attitude to IPV and barriers to health care variables. Women living in countries/time spans in which women held higher proportions of seats in parliament were less likely to believe that IPV was justified if the woman refuses to have sex compared to those living in settings in which women held fewer or no seats in parliament. And women living in countries/time spans in which women held higher proportions of seats in parliament were less likely to face financial or distance barriers to obtaining their own health care, over and above other effects, including the effect of the aggregate WGI governance indicator (results not shown).

CPIA gender equality as an alternative measure

Given that the aggregate WGI good governance indicator did not provide significant results for all independent variables and that it often operated in the opposite direction, meaning that better governance, according to the World Bank, was related to *fewer* women who easily read passages and more women *agreeing* that IPV is justified if women refuse to have sex, I thought it important to examine women-focused proxies of good governance as the sole governance variable. My hypothesis was that these would be better-fitting models when examining outcomes for a dataset of women in three countries, each at two distinct time periods. The variable I chose to focus on is the aggregate variable for women's equality, the CPIA gender equality variable (as defined earlier in this chapter).

The results of this analysis can be found in Table 5.5. I do not repeat the results from models 1–3 and 5. In Table 5.5, models 4 and 6 include the women-focused contextual variable that is not captured in CPIA gender equality, the proportion of seats in parliament held by women (repeated for convenience from model 4; see Table 5.4), and CPIA gender equality (model 6).

CPIA gender equality and women's outcomes

The results from model 4 shown in Table 5.5 are described above. Turning to model 6, which adds CPIA gender equality to model 4, in relation to respondent's literacy, women are more likely to read a passage easily if they live in urban areas and in later years of the survey. Women respondents are two times more likely to read a passage easily if they live in Tanzania compared to Zimbabwe. Women in households with a cell phone are 1.6 times more likely to read passages easily compared to women living in households without cell phones. Knowledge diffusion and gender equality variables do not yield consistent or expected results in literacy models.

Women respondents' attitudes toward IPV are related to the CPIA gender equality rating. Women living in countries and during survey years in which their country has a higher CPIA gender equality rating are only .802 times more likely to agree that "wife beating is justified" if women refuse to have sex. The control, ICT, and contextual variables operate as expected, and are consistent with the results discussed from model 4.

Table 5.5a Analysis of models 4 and 6: control factors, ICT, knowledge diffusion, gender-inclusive governance, and aggregate CPIA governance indicator as determinants of women's outcomes in Malawi, Tanzania, and Zimbabwe (2010–2011 and 2015–2016)

Dependent variable: literacy	*Gender-inclusive-governance and cannot read passage*		*CPIA governance and cannot read passage*		*Gender-inclusive-governance and reads passage easily*		*CPIA governance and reads passage easily*	
	Model 4		*Model 6*		*Model 4*		*Model 6*	
	Exp(B)	*Sig.*	*Exp(B)*	*Sig.*	*Exp(B)*	*Sig.*	*Exp(B)*	*Sig.*
Control variables								
Urban residence	0.703	***	0.703	***	1.355	***	1.354	***
Survey year	0.739	***	0.888	***	1.323	***	1.088	***
Country: Malawi	1.151	***	1.977	***	1.394	***	0.952	
Country: Tanzania	0.785	***	0.871	*	2.431	*	2.116	***
Country: Zimbabwe (omitted category)								
Resident is a visitor (versus usual resident)	1.001	***	1.001		0.946	***	0.945	
Female respondent is household head	1.092	***	1.092		1.006	***	1.005	
Proportion of heads of household who are male	1.162	***	1.162	***	0.927	***	0.928	*
Relationship structure in household, 2 adults of opposite sex	1.065	***	1.066		0.937	***	0.935	*
Relationship structure in household, 3+ related adults	0.933	***	0.934	*	1.048	***	1.048	
Woman is currently married	1.058	***	1.056	*	0.965	***	0.97	
Age of female respondent	1.011	***	1.011	***	0.987	***	0.987	***
ICT								
Household has cell phone	0.549	***	0.549	***	1.637	***	1.638	***

(*continued*)

Table 5.5a (cont.)

Dependent variable: literacy	*Gender-inclusive-governance and cannot read passage*		*CPIA governance and cannot read passage*		*Gender-inclusive-governance and reads passage easily*		*CPIA governance and reads passage easily*	
	Model 4		*Model 6*		*Model 4*		*Model 6*	
	Exp(B)	*Sig.*	*Exp(B)*	*Sig.*	*Exp(B)*	*Sig.*	*Exp(B)*	*Sig.*
Knowledge diffusion								
Female respondent, did not receive formal education	0.017	***	0.017	***	65.911	***	65.671	***
Female respondent, primary educational level (omitted category)								
Female respondent, secondary or further education	0.029	***	0.029	***	19.53	***	19.719	***
Contextual factors								
Proportion of seats held by women in national parliaments	0.976	***	1.025	***	1.022	***	0.987	**
Country policy and institutional assessment: gender equality								
CPIA gender equality rating (1=low to 6=high)			0.923				0.878	*
Constant	108.021	***	7.622	***	0.004	***	0.061	***

Source: Zerai's analysis of data extract performed by DHS implementing partners and ICF International. DHS: Malawi 2010, 2015–2016; Tanzania 2010, 2015–2016; Zimbabwe 2010, 2015–2016. Data extract from DHS Recode files. IPUMS-Demographic and Health Surveys (IPUMS-DHS), version 3.0, Minnesota Population Center and ICF International (Distributors). Accessed at: http://idhsdata.org

Notes:
B = coefficient; Sig. = significance level; Exp(B) = odds ratio.

Significance levels are grouped into three categories: *p<0.05; **p<0.01; ***p<0.001.

Table 5.5b Analysis of models 4 and 6: determinants of women's agreement that wife beating is justified

Dependent variable: female respondent agrees wife beating is justified if refuses sex	*Gender-inclusive-governance and respondent agrees wife beating is justified*		*CPIA governance and respondent agrees wife beating is justified*	
	Model 4		*Model 6*	
	Exp(B)	*Sig.*	*Exp(B)*	*Sig.*
Control variables				
Urban residence	0.639	***	0.64	***
Survey year	0.709	***	0.709;	***
Country: Malawi	0.225	***	0.256	***
Country: Tanzania	2.039	***	1.831	***
Country: Zimbabwe (omitted category)				
Resident is a visitor (versus usual resident)	1.149	*	1.151	**
Female respondent is Household head	0.987		0.988	
Proportion of heads of household who are male	1.063		1.062	
Relationship structure in HH, 2 adults of opposite sex	0.968		0.966	
Relationship structure in HH, 3+ related adults	1.006		1.009	
Woman is currently married	0.931	**	0.939	**
Age of female respondent	0.986	***	0.986	***
ICT				
Household has cell phone	0.811	***	0.807	***
Knowledge diffusion				
Female respondent, did not receive formal education	0.904	**	0.903	**
Female respondent, primary educational level (omitted category)				

(*continued*)

Table 5.5b (cont.)

Dependent variable: female respondent agrees wife beating is justified if refuses sex	*Gender-inclusive-governance and respondent agrees wife beating is justified*		*CPIA governance and respondent agrees wife beating is justified*	
	Model 4		*Model 6*	
	Exp(B)	*Sig.*	*Exp(B)*	*Sig.*
Female respondent, secondary or further educational level	0.483	***	0.484	***
Contextual factors				
Proportion of seats held by women in national parliaments	0.975	***	0.99	*
Country policy and institutional assessment: gender equality				
CPIA gender equality rating (1=low to 6=high)			0.802	***
Constant	1.382	***	1.978	***

Table 5.5c Analysis of models 4 and 6: determinants of women's money and distance barriers to medical care

Dependent variables: money and distance barriers to health care for respondent	*Gender-inclusive-governance and money a barrier to health care*		*CPIA governance and money a barrier to health care*		*Gender-inclusive-governance and distance a barrier to health care*		*CPIA governance and distance a barrier to health care*	
	Model 4		*Model 6*		*Model 4*		*Model 6*	
	Exp(B)	*Sig.*	*Exp(B)*	*Sig.*	*Exp(B)*	*Sig.*	*Exp(B)*	*Sig.*
Control variables								
Urban residence	0.603	***	0.605	***	0.271	***	0.272	***
Survey year	0.547	***	0.559	***	0.498	***	0.597	***

Country: Malawi	0.309	***	2.394	***	0.489	***	4.797	***
Country: Tanzania	0.23	***	0.161	***	0.231	***	0.186	***
Country: Zimbabwe (omitted category)								
Resident is a visitor (versus usual resident)	1.171	***	1.174	***	1.176	***	1.181	***
Female respondent is household head	1.044		1.042		1.073	*	1.071	*
Proportion of heads of household who are male	0.839	***	0.845	***	1.001		1.007	
Relationship structure in household, 2 adults of opposite sex	1.148	***	1.133	***	1.056	*	1.045	
Relationship structure in household, 3+ related adults	1.047	*	1.045	*	1.012		1.01	
Woman is currently married	0.756	***	0.775	***	0.969		0.991	
Age of female respondent	1.013	***	1.013	***	1.002	**	1.002	*
ICT								
Household has cell phone	0.646	***	0.627	***	0.748	***	0.737	***
Knowledge diffusion								
Female respondent, did not receive formal education	0.805	***	0.788	***	0.861	***	0.847	***
Female respondent, primary educational level (omitted category)								
Female respondent, secondary or further educational level	0.6	***	0.605	***	0.669	***	0.675	***
Contextual factors								
Proportion of seats held by women in national parliaments	0.948	***	1.156	***	0.934	***	1.16	***
Country policy and institutional assessment: gender equality								
CPIA gender equality rating (1=low to 6=high)			0.117	***			0.134	***
Constant	318.203	***	95.345	***	773.811	***	32.444	***

The results for model 6 that address barriers to women's health care are similar to those for model 4, with the exception of the fact that women living in Malawi are 2.394 times more likely to experience money as a barrier to their health care compared to Zimbabwe. And women living in Malawi are 4.797 times more likely to experience distance as a barrier to their health care compared to Zimbabwe. Also, the proportion of seats in parliament, as a proxy for women-centered political governance, changes direction in model 6 compared to model 4. So, women in countries in which women hold more seats in parliament are more likely to have financial and distance barriers to health care when CPIA gender equality is added to the model. Furthermore, model 6 shows that the higher the CPIA gender equality rating, the less likely women are to experience financial and distance barriers to health care.

The case for more inclusive measures of good governance

These results underline the necessity to broaden measures of good governance that capture dimensions that are important to women, for example, violence against women and whether IPV is prosecuted, women's political participation, including the proportion of seats held by women in parliament, safety net coverage, the education and literacy of women and girls, gender in nondiscrimination laws, and gender equality more generally.

Similar to Asongu and Nwachukwu (2016), I created a factor estimate for good governance in Malawi, Tanzania, and Zimbabwe in 2010 and 2014. The results show that this factor, which is an aggregate of political, economic, and institutional governance, sometimes has an impact on women's health outcomes. But models that include women-centered dimensions of good governance had more explanatory power, even when these variables were examined with control, ICT, and knowledge diffusion variables, and the WGI factor accounted for.

Notes

1 See Table 2.2 for a list of variables in the combined analysis file and data sources.

2 These data are from Daniel Kaufmann, Aart Kraay and Massimo Mastruzzi (accessed at: www.govindicators.org). I am using the version that was released in September 2017.

3 See https://databank.worldbank.org/data/reports.aspx?source=worldwide-governance-indicators#

4 Table 1.17 provides the percentages of women aged 15–49 who owned a cell phone in the latest Malawi, Tanzania, and Zimbabwe DHS.

5 See http://info.worldbank.org/governance/wgi/#home and https://cpia.afdb.org/documents/public/cpia2015-questionnaire-en.pdf for a fuller explanation.

6 CPIA data downloads analyzed in this manuscript are derived from the World Bank website (World Bank Group, CPIA database), accessed at: www.worldbank.org/ida

7 See http://info.worldbank.org/governance/wgi/#home and https://cpia.afdb.org/documents/public/cpia2015-questionnaire-en.pdf for a description of the six WGI indicators and the CPIA indicators

8 As in previous chapters, the literacy variable is operationalized utilizing an ordinal scale in the following way. A woman is literate if she has either completed secondary education or reads a passage easily. If she has not completed secondary education and cannot read a passage at all, she is not considered to be literate. If she has not completed secondary education and reads a passage with difficulty, she is considered partially literate. Women are asked to read passages in English or a mix of indigenous languages, depending upon the country in which the DHS is taking place.

9 The DHS collects data on all women in households and distinguishes between usual residents and visitors.

Conclusion

In this book, I have critically examined factors that the World Bank uses to describe good governance and called for an intervention by African women scholars. I have applied an African feminist lens to the study conducted by Asongu and Nwachukwu (2016) and found: (1) their model analyzing the impact of ICT on knowledge economies and good governance lacks factors that consider issues of gender equity in the definition of good governance; and (2) ICT access and knowledge diffusion must be measured in ways that consider women's access.

A more inclusive definition of good governance

In the first chapter of this book, I postulated that the World Bank's Worldwide Governance Indicators (WGI) project is incomplete, that it posits an androcentric perspective of development and governance. Throughout this book, I have recommended a number of factors that should be added that highlight women's access to, and women as beneficiaries of, governance. All of these factors are currently available in either national or cross-national datasets. As I have shown in this book's analyses, the CPIA in particular offers useful cross-national data, as do the Demographic and Health Surveys. I summarize my gender-sensitive recommendations below.

Political governance

1. For the political stability and absence of violence/terrorism dimension, I recommend adding violence against women in the last 12 months.
2. For the voice and accountability dimension, I recommend adding women's political activity, including women elected to parliament, and percentage of women voting (if available).

Economic governance

1. For regulatory quality, I recommend adding women's equal pay, and gender in nondiscrimination laws.
2. For government effectiveness, I recommend a number of variables: availability of safety nets, GPI, female access to primary school, and higher equity in access to water/sanitation.

Institutional governance

1. For control of corruption, I recommend adding whether a country has laws against sexual harassment.
2. For rule of law, I recommend IPV prosecution and percentage of women experiencing violence in the last 12 months.

In my intervention, I ask does adding variables that highlight women's access improve the predictive power of this framework for women and people generally? In Figure 2.3, I posit that it would be useful to examine cell phone use and its impact on education and knowledge diffusion among women.

Analyses relying on World Bank understandings of governance (such as Asongu and Nwachukwu 2016), do not consider evidence of women's involvement in the knowledge economy or governance factors that directly affect women. Asongu and Nwachukwu's results, then, largely explain *men's* access to better governance and knowledge diffusion. Examining the weaknesses of Asongu and Nwachukwu's model through the lens of African feminism and women scholars on the continent, influenced my development of a new, organic, gender-sensitive model of ICT and good governance.

While their analyses produced robust results, Asongu and Nwachukwu do not address gender asymmetry. For example, they do not consider women's access to ICT.

Some factors that I included that move beyond the analysis by Asongu and Nwachukwu (that inspired *this* work) are: the percentage of cell phones in women's households, women's mobile-cellular telephone subscriptions, and women's cell phone ownership. I further proposed that if ICT and knowledge diffusion have a positive impact on good governance (as shown in Asongu and Nwachukwu 2016), then governance should be good for everyone, including women, that is it should have a measureable impact on individual women's lives. I analyzed the impact of ICT on community level and national governance factors that include values of gender equity. I find a strong statistically significant relationship between ICT (cell phone ownership), knowledge diffusion, and *both* androcentric and gender-inclusive factors that either proxy good governance (e.g. government effectiveness and rule of law) or are results of good governance (e.g. women's literacy, educational attainment, and access to water/sanitation and health care). These findings are relevant to studies of science and technology in society, development effectiveness, public administration of ICT, and the study of women and gender in African countries.

In future work, it may be useful to consider the impact of Country Policy and Institutional Assessment (CPIA) variables developed by the African Development Bank Group, and especially the gender equality aggregate variable. "Women hold up half the sky" and we cannot fully understand what it means to have good governance if it is not also good for women's lives.

As more Demographic and Health Survey and other nationally representative household datasets become available, it would be useful to expand the analyses herein to examine differential impacts of ICT and knowledge diffusion among and between regions within the continent of Africa. A natural next step would be an examination of Southern and East Africa. A 16-country regional analysis[1] should include women's governance factors such as women elected to parliament, safety net coverage, GPI, female access to primary school, women's equal pay, gender in nondiscrimination laws, and IPV prosecutions. And a mixed methods design would make such an analysis even more powerful. Including qualitative assessments of whether material support of the Convention on the Elimination of All Forms of Discrimination against Women (CEDAW) exists, and the extent to which laws to promote gender equity are *actually carried out*, and that accountability measures are in place and work, would add significantly to such an analysis.

Currently, organizations such as the African Union, the Pan African Parliament (PAP), and a number of parliaments and other governing bodies on the African continent point to their gender policies (as described in Semu 2002, 2004). For example in 2009, the African Union adopted its Gender Policy "which enshrined the Gender Equality Principle, the Protocol to the African Charter on Human and Peoples' Rights on the Rights of Women in Africa, and the Solemn Declaration on Gender Equality in Africa" according to the then Chairperson of the African Union, H.E. Col. Muammar al Gaddafi (2009, p. ii). And for the past ten years, the African Union Commission has worked with the

Women, Gender and Development Directorate (WGDD) to organize annual Gender Pre-Summits. Results from this work have included:

> the adoption of the first AU Gender Policy in 2009; the establishment of the Fund for African Women (FAW) in 2010; the adoption of the African Women's Decade (AWD) 2010–2020; the appointment of the first African Union Commission Chairperson's Special Envoy on Women, Peace and Security in 2013 and the Declaration of the AU Heads of State and Government of 2015 as the Year of Women's Empowerment and Development Towards Agenda 2063, and 2016 as the Year on Human Rights, with a special focus on the rights of women in Africa.
>
> (African Union 2018, n.p.)

These efforts are laudable; but realizing the goal of gender equity requires education to gain awareness of implicit bias, as well as institutional transformation to accomplish appropriate cultural shifts to make governance and governing bodies, in particular, welcoming to women.

In a blog written by Irene Kagoya of the feminist organization, Akina Mama Wa Afrika, she indicates her surprise to witness how inhospitable the Ugandan Parliament was to women. This unwelcoming climate was described as physical violence that took place when elected officials did not agree. Kagoya remembers:

> While the session kicked off well, the tension in the house was evident from both the opposition and ruling party. It was not long before a scuffle ensued between one opposition member and a member of the ruling party. Members begun lifting chairs and exchanging blows. It was complete chaos but amidst all this, Honourable Monica Amoding stood strong and held her ground. She even tried to stop the fighting at one point. As I watched the drama, my mind kept racing – thinking about women and effective political participation. How can women effectively engage in political leadership when their safety is threatened? What does this mean for any young woman wishing to join political leadership? In no circumstance would this environment entice women to participate in governance and democratic processes.

Writing for Akina Mama Wa Afrika, Kagoya reiterates a point made by Professor Semu and others, that nominally providing seats to women in parliament is not enough. Parliaments must not only be more diverse, in terms of gender, but in order to realize the full benefit of women's membership, they must become *inclusive*. Kagoya continues:

> For a long time we have continued to question women's meaningful participation in Parliament. While the rise of women in parliament with statistics at 35% has not substantially translated into the change we seek; we must remain cognisant of the environment within which these women operate. This for me was a take home. It is not enough to have them in the room; we must support them to fully influence the decisions in parliament. Part of this is advocating for a safe environment, including procedural rules that deliberately position women to engage in the parliamentary session.[2]

And Ms. Kagoya further noted that the poor climate was perpetuated by body shaming against women parliament members. And she analyzes the motivation for this cruel behavior:

> When I talk of a safe environment I mean an environment where women are free, and treated on equal footing as their male counterparts in all rights. It was painful to watch one opposition member raise a point of order on a matter of a female member's dressing stating that she dresses indecently and violates him, a matter that the speaker ruled on. For me this was yet again another tactic of policing women's bodies. It was evident that he wanted to embarrass her so that she is discouraged from confidently pursuing her agenda as a leader of the ruling party which was clearly motivated by his egoism. As I kept watching, I could also see one female member fidgeting with her skirt slit throughout the session as her male counterparts playing out in a parliament on that day. We ought to ensure that all MPs regardless of their differences are accorded the same respect and dignity as provided in the 1995 constitution.

Successfully accomplishing the "gender equality principle" is not a one-time act. Education for critical consciousness is at the base of it, for women and for men. And men who are currently enjoying the privileges of patriarchy need to be willing to relinquish some power. Additionally, older women who think they benefit from patriarchy need to be willing to relinquish power and teach the youth in new ways. Institutional transformation to promote gender equity must address corruption. Corruption impedes equity (as recognized in the recent Tenth African Union Gender Pre-Summit to Discuss Impact of Corruption on

Women's Empowerment). Nepotism impedes equity. Favoritism to particular ethnic groups impedes equity. Voting/elections that are not free or fair impede equity. Violence to resolve disputes impedes equity. To the extent that these impediments exist, efforts to increase diversity will not accomplish true inclusion of women's voices and concerns in governance.

In future work, I seek to continue to build the African feminist scholar/activist database. I would like to work with the journals such as *Jenda: A Journal of Culture and African Women Studies*, and *Feminist Africa* (founded by Professor Amina Mama); and with The Council for the Development of Social Science Research in Africa (CODESRIA), the Africa Knowledge Project (founded by SUNY Binghamton's Professor Nkiru Nzegwu), and other women-centered pan-African organs that are currently at the forefront of articulating women's concerns and who should be called upon to develop what it means to have good governance and how to democratize access to ICT. *A luta continua!*

Notes

1 The analysis is proposed to include: Angola, Burundi, Botswana, Kenya, Lesotho, Mozambique, Malawi, Namibia, Rwanda, South Sudan, Swaziland, Tanzania, Uganda, South Africa, Zambia, and Zimbabwe.

2 This undated entry can be accessed at: www.akinamamawaafrika.org/index.php/blog/199-a-feminist-lense-to-uganda-srecent-parliamentary-ordeal.

References

ActionAid. 2005. An ActionAid and Ayuda en Acción Policy Briefing for the World Conference on Disaster Reduction: To Achieve a Safer World for Poor and Excluded People, Disaster Reduction Must be Underpinned by People-Centered Governance. Japan, January 18–22, 2005. Accessed at: www.actionaid.org/sites/files/actionaid/people-centred_governance_reducing_disaster_for_poor_and_excluded_people.pdf

African Development World Bank. 2015. *The Country Policy and Institutional Assessment of the African Development Bank 2015 Questionnaire and Guidelines*. Accessed at: https://cpia.afdb.org/documents/public/cpia2015-questionnaire-en.pdf

African Union. 2009. *Gender Policy*. Accessed at: www.un.org/en/africa/osaa/pdf/au/gender_policy_2009.pdf

Agence Nationale de la Statistique et de la Démographie (ANSD) [Senegal] and ICF International. 2015. *Sénégal: Enquête Continue sur la Prestation des Services de Soins de Santé (ECPSS) 2014*. Rockville, MD: ANSD and ICF International.

Agozino, Biko. 1999. "Committed Objectivity in Race-Class-Gender Research." *Quality and Quantity*, 33(4), 395–410.

Agozino, Biko. 2007. "Intrusive Traits of Victor Chikezie Uchendu." *Dialectical Anthropology*, 31(13), 289–92.

Amadiume, Ifi. 1987. *Male Daughters, Female Husbands: Gender and Sex in African Society*. London: Zed Books.

Amadiume, Ifi. 1997. *Re-inventing Africa: Matriarchy, Religion and Culture*. London: Zed Books.

Amadiume, Ifi. 2000. *Daughters of the Goddess, Daughters of Imperialism: African Women Struggle for Culture, Power and Democracy*. Available from *Historical Abstracts*, Ipswich, MA.

Amnesty International. 2008. Interview with WOZA co-founder and executive director Jenni Williams, recorded after her release from prison. Conducted by Rolly Brook, country specialist for Amnesty International USA, and colleagues. Accessed at: http://luxmedia.com.edgesuite.net/amnesty/wozainterview.mp3

Amundsen, I. and Happy Kayuni (eds). 2016. *Women in Politics in Malawi*. Bergen and Zomba: Chr. Michelsen Institute (CMI), Department of Political and Administrative Studies.

Asongu, S. and J. Nwachukwu. 2016. "The Mobile Phone in the Diffusion of Knowledge for Institutional Quality in Sub-Saharan Africa." *World Development*, 86, 133–47.

Barnes, Teresa. 1995. "The Heroes' Struggle: Life after the Liberation War for Four ExCombatants." In N. Bhebe and T.O. Ranger (eds), *Soldiers in Zimbabwe's Liberation War*, pp. 18–138. Harare: University of Zimbabwe Publications.

Barnes, Teresa. 1999. *"We Women Worked So Hard": Gender, Urbanization, and Social Reproduction in Colonial Harare, Zimbabwe, 1930–1956*. Portsmouth, NH: Heinemann.

Barnes, Teresa. 2002. "Virgin Territory? Travel and Migration by African Women in Twentieth-Century Southern Africa." In S. Geiger, N. Musisi, and J.M. Allman (eds), *Women in African Colonial Histories*, pp. 164–90. Bloomington, IN: Indiana University Press.

Barnes, Terri and Everjoyce Win. 1992. *To Live a Better Life: An Oral History of Women in the City of Harare, 1930–70*. Harare: Baobab Books.

Bayat, Asef. 2012. "Politics in the City-Inside-Out." *City and Society*, 24(2), 110–28.

Bazargan-Hejazi, S., S. Medeiros, R. Mohammadi, J. Lin, and K. Dalal. 2013. "Patterns of Intimate Partner Violence: A Study of Female Victims in Malawi." *Journal of International Violence Research*, 5(1), 38–50.

Bell, D.A. 1980. "Brown v. Board of Education and the Interest-Convergence Dilemma." *Harvard Law Review*, 93(3), 518–33.

Bowen, Merle and Ayesha S. Tillman. 2014. "Developing Culturally Responsive Surveys: Lessons in Development, Implementation, and Analysis from Brazil's African Descent Communities." *American Journal of Evaluation*, July 2, 1–17.

Bratton, Michael. 1993. "Reluctant Rebels Review of Zimbabwe's Guerrilla War: Peasant Voices." *Journal of African History*, 34, 163–5.

Brautigam, Deborah. 2009. *The Dragon's Gift: The Real Story of China in Africa*. Oxford: Oxford University Press.

Brewer, Rose. 1992. "Black Studies Transformed: A Consideration of Gender, Feminism, and Black Women's Studies." In C. Kramarae and D. Spender (eds), *The Knowledge Explosion: Generations of Feminist Scholarship*, pp. 64–70. New York, NY: Teachers' College Press.

Brewer, Rose. 1993. "Theorizing Race, Class, and Gender: The New Scholarship of Black Feminist Intellectuals and Black Women's Labor." In S.M. James and A. Busia (eds), *Theorizing Black Feminism: The Visionary Pragmatism of Black Women*, pp. 13–30. London: Routledge.

Brewer, Rose. 1999. "Theorizing Race, Class, and Gender: The New Scholarship of Black Feminist Intellectuals and Black Women's Labor." *Race, Class and Gender: An Interdisciplinary and Multidisciplinary Journal*, 6(2), 29–47.

Campbell, Horace. 2003. *Reclaiming Zimbabwe: The Exhaustion of the Patriarchal Model of Liberation*. Claremont, South Africa: David Philip.

Caulderwood, Kathleen. 2015. Connections in Sub-Saharan Africa Will Spur Development: World Bank Economist. *International Business Times*. March 20. Accessed at: www.ibtimes.com/mobile-connections-subsaharan-africa-will-spur-development-world-bank-economist-1853960

Chaputula, Aubrey Harvey and Stephen Mutula. 2018. "Provision of Library and Information Services through Mobile Phones in Public University Libraries in Malawi." *Global Knowledge, Memory and Communication*, 67(1/2), 52–69.

Charmaz, Kathy and Debora A. Paterniti. 1999. "Introduction." In *Health, Illness and Healing: Society, Social Context, and Self*, pp. 5–30. Los Angeles, CA: Roxbury Publishing.

Chawinga, Winner and F. Ngwira. 2015. "The Role of the Mzuzu American Corner Information Technology and Communication Centre, Malawi." *Innovation: Journal of Appropriate Librarianship and Information Work in Southern Africa*, 50, 20–23.

Chikasanda, Vanwyk Khobidi Mbubzi, Kathrin Otrel-Cass, John Williams, and Alister Jones. 2013. "Enhancing Teachers' Technological Pedagogical Knowledge and Practices: A Professional Development Model for Technology teachers in Malawi." *International Journal of Technology and Design Education*, 23(3), 597–622.

Chikasanda, Vanwyk Khobidi, Mgawi, Rabson Killion, Doris Mtemang'ombe, and Yusuf Alide. 2015. "Introducing Technology Studies in Malawi's Model Primary Schools: Towards Building a Technologically Literate Society." *International Journal of Technology and Design Education*, 25(4), 453–66.

Chilimo, Wanyenda Leonard. 2010. *Information and Communication Technologies and Sustainable Livelihoods: A Case of Selected Rural Areas of Tanzania*. Networked Digital Library of Theses & Dissertations.

Chima, Angeline Liza Ndachiphata and Vanwyk Khobidi Chikasanda. 2013. "The Impact of Internet Banking on Service Quality Provided by Commercial Banks." In e-Infrastructure and e-Services for Developing Countries, 5th International Conference, AFRICOMM 2013, Blantyre, Malawi, November 25–27, 2013, Revised Selected Papers 135, pp. 248–59. Springer International Publishing.

Chinakidzwa, More, Pinigas Mbengo, and Marcelene Nyatsambo. 2015. "Mobile Money Usage in Rural Areas of Zimbabwe – Case of Mudzi District." *International Journal of Scientific & Engineering Research*, 6(2), n.p.

Chingwete, Anyway, Samantha Richmond, and Carmen Alpin. 2014. "Support for African Women's Equality Rises." *Afrobarometer*. March 27.

Chipfuwa, T., A. Manwere, M.M. Kuchenga, L. Makuyana, E. Mwanza, E. Makado, and R.P. Chimutso. 2014. "Level of Awareness and Uptake of the Female Condom in Women aged 18 to 49 Years in Bindura District, Mashonaland Central Province, Zimbabwe." *African Journal of AIDS Research*, 13(1), 75–80.

Chunga, Richard M., Jeroen H.J. Ensink, Marion W. Jenkins, and Joe Brown. 2016. “Adopt or Adapt: Sanitation Technology Choices in Urbanizing Malawi.” *PLoS ONE*, 11(8), e0161262. doi:10.1371/journal.pone.0161262.

Clarke, John Henrik. 1974. “Cheikh Anta Diop and the New Light on African History.” Archives of *FRONTal View: An Electronic Journal of African Centered Thought*. Accessed at: www.nbufront.org/MastersMuseums/JHClarke/Contemporaries/CheikhAntaDiop.html

Clarke, John Henrik. 1990. Introduction to *The Cultural Unity of Black Africa: The Domains of Patriarchy and of Matriarchy in Classical Antiquity* by Cheikh Anta Diop. Chicago, IL: Third World Press.

Collins, Patricia Hill. 1990. *Black Feminist Thought: Knowledge, Consciousness, and the Politics of Empowerment*, 2nd edn. Boston, MA: Unwin Hyman.

Collins, Patricia Hill. 1998a. *Fighting Words: Black Women and the Search for Justice*. Minneapolis: University of Minnesota Press.

Collins, Patricia Hill. 1998b. “Intersections of Race, Class, Gender, and Nation: Some Implications for Black Family Studies.” *Journal of Comparative Family Studies* 29 (Spring), 27–36.

Conrad, Peter and Valerie Leiter (eds). 2003. “Introduction.” In *Health and Health Care as Social Problems*, pp. 1–6. Lanham, MD: Rowman & Littlefield.

Cools, Sara and Andreas Kotsadam. 2017. “Resources and Intimate Partner Violence in Sub-Saharan Africa.” *World Development*, 95, 211–30.

Cooper, Anna Julia. 1998. *The Voice of Anna Julia Cooper: Including A Voice From the South and Other Important Essays, Papers, and Letters* (Charles Lemert and Esme Bhan, eds). Lanham, MD: Rowman & Littlefield.

Crawford, Jessica, Erin Larsen-Cooper, Zachariah Jezman, Stacey C. Cunningham, and Emily Bancroft. 2014. “SMS versus Voice Messaging to Deliver MNCH Communication in Rural Malawi: Assessment of Delivery Success and User Experience.” *Global Health: Science and Practice*. January 2014, ghs1300155; https://doi.org/10.9745/GHSP-D-13-00155.

Daley, Patricia. 2008. *Gender and Genocide in Burundi: The Search for Spaces of Peace in the Great Lakes Region*. Oxford: James Currey.

Daley, Patricia. 2014. “Researching Sexual Violence in the Eastern Democratic Republic of Congo: Methodologies, Ethics, and the Production of Knowledge in an African Warscape.” In A. Coles, L. Gray, and K. Momsen (eds), *Routledge Handbook of Gender and Development*, pp. 429–40. New York, NY: Routledge.

Davies, Rob. 2004. “Memories of Underdevelopment: A Personal Interpretation of Zimbabwe’s Economic Decline.” In B. Raftopoulos and T. Savage (eds), *Zimbabwe: Injustice and Political Reconciliation Institute for Justice and Reconciliation (South Africa)*, pp. 1–18. Harare: Weaver Press.

DeLaet, D. and D. DeLaet. 2012. *Global Health in the 21st Century: The Globalization of Disease and Wellness*. Boulder, CO: Paradigm Publishers.

Demirguc-Kunt, Asli, Leora Klapper, Dorothe Singer and Peter Van Oudheusden. 2015. Global Financial Inclusion Database. World Bank; International Telecommunication Union, World Telecommunication/ICT Development Report and database.

Denzin, N.K. and Y.S. Lincoln. 2008. "Introduction: Critical Methodologies and Indigenous Inquiry." In N.K. Denzin, Y.S. Lincoln and L. Tuhiwai Smith (eds), *Handbook of Critical and Indigenous Methodologies*, pp. 1–20. Thousand Oaks, CA: SAGE Publications.

Devault, Marjorie. 1991. *Feeding the Family: The Social Organization of Caring as Gendered Work*. Chicago, IL: University of Chicago Press.

Devault, Marjorie. 1999. *Liberating Method: Feminism and Social Research*. Philadelphia, PA: Temple University Press.

Devries, K.M., J.Y. Mak, C. Garća-Moreno, M. Petzold, J.C. Child, G. Falder, S. Lim, L.J. Bacchus, R.E. Engell, L. Rosenfeld, C. Pallitto, T. Vos, N. Abrahams, and C.H. Watts. 2013. "The Global Prevalence of Intimate Partner Violence against Women." *Science*, 340(6140), 1527–8.

DHS implementing partners and ICF International. Demographic and Health Surveys. 2004–2014 [Egypt, 2005, 2008, 2014; Ethiopia 2005; Kenya 2008–2009; Malawi 2004, 2010; Mozambique 2011; Nigeria 2008, 2013; Tanzania 2010; Uganda 2006, 2011; Zambia 2007, 2013; Zimbabwe 2005–2006, 2010]. Data extract from DHS Recode files. IPUMS Demographic and Health Surveys (IPUMS-DHS), version 3.0, Minnesota Population Center and ICF International [Distributors]. Accessed at: http://idhsdata.org

Diamond, Timothy. 1992. *Making Gray Gold: Narratives of Nursing Home Care*. Chicago, IL: University of Chicago Press.

Dillard, Cynthia B. 2000. "The Substance of Things Hoped For, The Evidence of Things Not Seen: Examining an Endarkened Feminist Epistemology in Educational Research and Leadership." *Qualitative Studies in Education*, 13(6), 661–81.

Dillard, Cynthia B. and Chinwe Okpalaoka. 2011. "The Sacred and Spiritual Nature of Endarkened Transnational Feminist Praxis in Qualitative Research." In N.K. Denzin and Y.S. Lincoln (eds), *The SAGE Handbook of Qualitative Research*, pp. 147–62. Thousand Oaks, CA: SAGE Publications.

Dillard, Cynthia B., Daa'iyah Abdur-Rashid, and Cynthia A. Tyson. 2000. "My Soul is a Witness: Affirming Pedagogies of the Spirit." *Qualitative Studies in Education*, 13(5): 447–62.

Du Bois, W.E.B. 1897. "A Program for a Sociological Society" [Unpublished Manuscript] W.E.B. Du Bois Papers, Department of Special Collections and University Archives, W.E.B. Du Bois Library, University of Massachusetts at Amherst.

Du Bois, W.E.B. 1899. *The Philadelphia Negro: A Social Study*. Philadelphia, PA: University of Penn Pres.

Du Bois, W.E.B. 1903a. "Sociology Hesitant" [Unpublished manuscript]. W.E.B. Du Bois Papers, Department of Special Collections and University Archives, W.E.B.

Du Bois, W.E.B. 1903b. *The Souls of Black Folk*. Chicago, IL: A.C. McClurg & Co.

Fanon, F. 1963. *The Wretched of the Earth*. New York, NY: Grove Press.

Gaidzanwa, Rudo B. 1991. "Labour Relations in a Mining Enterprise Established After Independence." *Zambezia: The Journal of the University of Zimbabwe* 18(1), 49–67.

Gaidzanwa, Rudo B. 1992. "The Ideology of Domesticity and the Struggles of Women Workers: The Case of Zimbabwe." *Working Paper Series – Sub-series on Women, History and Development: Themes and Issues*, Institute of Social Studies, The Hague.

Gaidzanwa, Rudo B. 1993. "The Politics of Body and the Politics of Control: An Analysis of Class, Gender and Cultural Issues in Student Politics at the University of Zimbabwe." *Zambezia: The Journal of the University of Zimbabwe*, 20(1), 15–33.

Gaidzanwa, Rudo. 1994. "Women's Land Rights in Zimbabwe." *Issue: A Journal of Opinion*, 22(2), 12–16.

Gaidzanwa, Rudo. 1996. "The Ideology of Domesticity and the Struggles of Women Workers in Zimbabwe." In A. Chhachhi and R. Pittin (eds), *Confronting State Capital and Patriarchy*, pp. 273–89. New York, NY: St. Martin's Press.

Gaidzanwa, Rudo B. 1997. "Gender Analysis in the Field of Education: A Zimbabwean Example." In A. Imam, A. Mama, and F. Sow (eds), *Engendering African Social Sciences*, pp. 271–96. Dakar: CODESRIA.

Gaidzanwa, Rudo B. (ed.). 1998. "Policy Making Issues in Southern Africa." *Southern Africa Regional Institute for Policy Studies*. Mt. Pleasant, Harare, Zimbabwe: SAPES Books.

Gaidzanwa, Rudo B. 1999a. *Voting with Their Feet: Migrant Zimbabwean Nurses and Doctors in the Era of Structural Adjustment*. Uppsala, Sweden: Nordiska Afrikainstitutet.

Gaidzanwa, Rudo B. 1999b. "Indigenisation as Empowerment? Gender and Race in the Empowerment Discourse in Zimbabwe." In A. Cheater (ed.), *The Anthropology of Power: Empowerment and Disempowerment in Changing Structures*, pp. 118–32. New York, NY: Routledge.

Gaidzanwa, Rudo B. 2006. "Continuity and Change in Women's Shona and Ndebele Writing in Zimbabwe: A Gender Analysis." In Z. Mguni, M. Furusa, and R. Magosvongwe (eds), *African Womanhood in Zimbabwean Literature: New Critical Perspectives on Women's Literature in African Languages*, pp. 195–211. Harare: College Press.

Gaidzanwa, Rudo B. 2008. "Poverty, HIV and AIDS: Implications for Southern Africa in the Global Capitalist Architecture." *South-South Collaborative Programme Occasional Paper Series* #23.

Gaidzanwa, Rudo B. 2011. "Women and Land in Zimbabwe." Paper presented at the Conference: Why Women Matter in Agriculture. Stockholm, Sweden, April 4–8, 2011.

Gaidzanwa, Rudo B. 2013. "African Feminism." *BUWA! A Journal of African Women's Experiences*. Open Society Initiative for Southern Africa. Accessed at: www.osisa.org/buwa/regional/african-feminism-rudo-gaidzanwa

Gama, Mavis. 2013. "Gender Disparity in Zimbabwe Parliament Upsets Women." *VOA Zimbabwe*, September 5. Accessed at: www.voazimbabwe.com/a/zimbabwewomen-parliament-of-zimbabwe-proportional-representation/1743760.html

Garćia-Moreno, Jansen and Watts Heise Ellsberg. 2005. "WHO Multi-Country Study on Women's Health and Domestic Violence against Women: Initial Results on Prevalence, Health Outcomes and Women's Responses." Technical report. World Health Organization, Geneva.

Gengiah, Tanuja N., Michele Upfold, Anushka Naidoo, Leila E. Mansoor, Paul J. Feldblum, Quarraisha Abdool Karim, and Salim S. Abdool Karim. 2014. "Monitoring Microbicide Gel Use with Real-Time Notification of the Container's Opening Events." *AIDS Behavior* , 18, 833–40.

Giné, Xavier and Dean Yang. 2009. "Insurance, Credit, and Technology Adoption: Field Experimental Evidence from Malawi." *Journal of Development Economics*, 89(1), 1–11.

Gordon, L.R. 2006. *Disciplinary Decadence: Living Thought in Trying Times*. New York, NY: Paradigm Publishers.

GSMA. 2014. *Connected Women Striving and Surviving in Papua New Guinea: Exploring the Lives of Women at the Base of the Pyramid*. Accessed at: www.gsma.com/mobilefordevelopment/wpcontent/uploads/2014/11/mWomen_PN G_v3.pdf

GSMA Connected Women Global Development Alliancc. 2015. *Bridging the Gender Gap: Mobile Access in Low and Middle-Income Countries*.

GSMA, World Bank Data, and Altai Consulting Analysis. Accessed at: www.gsmamobileeconomy.com/GSMA_ME_Report_2014_R2_WEB.pdf

GSMA Intelligence 2014; 2015; 2016; 2017; 2018. The Mobile Economy. 205. Accessed at: www.gsma.com/mobileeconomy/

GSMA *The Mobile Economy: SubSaharan Africa* 2017.

Growth with Equity: An Economic Policy Statement. 1981. Harare: Government of the Republic of Zimbabwe.

Guillaumin, Colette. 1995. *Racism, Sexism, Power, and Ideology*. New York, NY: Routledge.

Gwede, Wanga. 2017. "Some Women Feel Husbands Justified in Wife Beating – Malawi Demographic and Health Survey." *Nyasa Times*. Accessed at: www.nyasatimes.com/women-feelhusbands-justified-wife-beating-malawi-demographic-health-survey/

Harding, Sandra and Kathryn Norberg. 2005. "New Feminist Approaches to Social Science Methodologies: An Introduction." *Signs*, 30(4), 2009–15.

Harrison, Faye. 2016. Personal Communication.

Henderson, Clara and Lisa Gilman. 2004. "Women as Religious and Political Praise Singers within African Institutions: The Case of the CCAP Blantyre Synod and Political Parties in Malawi." *Women and Music: A Journal of Gender and Culture*, 8, 22–40.

Henderson, Loren, A. Zerai, and R. Morrow. 2017. "Intimate Partner Violence and HIV Status among Ever-Married and Cohabiting Zimbabwean

Women: An Examination of Partners' Traits." *African Journal of Reproductive Health*, 21(4), 45–54.

hooks, bell and Cornel West. 1991. *Breaking Bread: Insurgent Black Intellectual Life*. Boston, MA: South End Press.

Huggins, C. and A. Valverde. 2018. "Information Technology Approaches to Agriculture and Nutrition in the Developing World: A Systems Theory Analysis of the Nutrition Program in Malawi." *Food Security*, 10(1), 151–68.

Human Rights Watch. 2017. "'I Had a Dream to Finish School': Barriers to Secondary Education in Tanzania." Human Rights Watch. February 2017 Issue. Accessed at: www.hrw.org/sites/default/files/accessible_document/tanzania0217__accessible.pdf

IDHS IPUMS-DHS implementing partners and ICF International. Demographic and Health Surveys Malawi 2010, 2016; Tanzania 2010, 2015–2016; Zimbabwe 2010, 2015–2016. Data extract from DHS Recode files. IPUMS-Demographic and Health Surveys (IPUMS-DHS), version 3.0, Minnesota Population Center and ICF International [Distributors]. Accessed at: http://idhsdata.org

IDHS IPUMS-DHS implementing partners and ICF International. Demographic and Health Surveys Egypt, 2005, 2008, 2014; Ethiopia 2005; Kenya 2008–2009, 2014; Malawi 2004, 2010; Mozambique 2011; Nigeria 2008, 2013, 2015; Tanzania 2010, 2015–2016; Uganda 2006, 2011, 2014–2015; Zambia 2007, 2013; Zimbabwe 2005–2006, 2010, 2015–2016. Data extract from DHS Recode files. IPUMS-Demographic and Health Surveys (IPUMS-DHS), version 3.0, Minnesota Population Center and ICF International [Distributors]. Accessed at: http://idhsdata.org

Imam, Ayesha. 1997. "An Introductory Essay." In A. Imam, A. Mama, and F. Sow (eds), *Engendering African Social Sciences*, pp. 1–31. Dakar: CODESRIA.

Imam, Ayesha M. and Amina Mama. 1994. "Role of Academics in Limiting and Expanding Academic Freedom." In M. Mamdani and M. Diouf (eds), *Academic Freedom in Africa*, pp. 73–107. Dakar: CODESRIA.

Imam, Ayesha M., Amina Mama, and Fatou Sow (eds). 1997. *Engendering African Social Sciences*. Dakar: CODESRIA.

Isaya, Elizabeth L., Robert Agunga, and Camilius A. Sanga. 2018. "Sources of Agricultural Information for Women Farmers in Tanzania." *Information Development*, 34(1), 77–89.

Jaji, Rosemary. 2009. "Masculinity on Unstable Ground: Young Refugee Men in Nairobi, Kenya." *Journal of Refugee Studies*, 22(2), 177–94.

Jakobsen, Hilde. 2016. "How Violence Constitutes Order: Consent, Coercion, and Censure in Tanzania." *Violence Against Women*, 24(1), 45–65.

Jerie, Steven. 2011. "Gender and Solid Waste Management in the Informal Sector of Bulawayo, Zimbabwe." *The Dyke*, 5(1), 46–64.

Jewkes, R., K. Dunkle, M. Nduna, and S. Nwabisa. 2010. "Intimate Partner Violence, Relationship Power Inequity, and Incidence of HIV Infection in Young Women in South Africa: A Cohort Study." *The Lancet*, 376(9734), 41–8.

Jones, Rachel. 2008. "Soap Opera Video on Handheld Computers to Reduce Young Urban Women's HIV Sex Risk." *AIDS Behavior* , 12, 876–84.
Kabote, Samwel J., Elliott P. Niboye, and Carolyne I. Nombo. 2014. "Performance in Mathematics and Science Subjects: A Gender Perspective for Selected Primary Schools in Rural and Urban Tanzania." *International Journal of Gender Studies and Women's Studies*, 5(3), 87–105.
Kahamba, J.S. and A.S. Sife. 2017. "Analysis of Gender Stereotypes in Television Advertisements in Tanzania." *University of Dar es Salaam Library Journal*, 12(2), 62–79.
Kakota, Tasokwa, Dickson Nyariki, David Mkwambisi and Wambui Kogi-Makau. 2011. "Gender Vulnerability to Climate Variability and Household Food Insecurity." *Climate and Development*, 3(4), 398–409.
Kagoya, Irene. n.d. A Feminist Lens to Uganda's Recent Parliamentary Ordeal. Accessed at: www.akinamamawaafrika.org/index.php/blog/199-a-feminist-lense-to-uganda-srecent-parliamentary-ordeal
Kaufman, M.R., R.N. Rimal, M. Carrasco, O. Fajobi, A. Soko, R. Limaye, and G. Mkandawire. 2014. "Using Social and Behavior Change Communication to Increase HIV Testing and Condom Use: The Malawi BRIDGE Project." *AIDS Care*, 6 Suppl 1, S46–9. doi: 10.1080/09540121.2014.906741.
Kaufmann, Daniel, Aart Kraay, and Massimo Mastruzzi. 2010. *The Worldwide Governance Indicators: Methodology and Analytical Issues*. World Bank, September. Accessed at: www.govindicators.org (note: version I am using is from September 22, 2017).
Kayuni, Happy M. and Kondwani Farai Chikadza. 2016. *The Gatekeepers: Political Participation of Women in Malawi*. Bergen: Chr. Michelsen Institute.
Kisanga, M. and N. Katunzi 1997. *Discrimination in Education The Tanzania Case Dar es Salaam*. UNESCO. Accessed at: http://unesdoc.unesco.org/images/0013/001385/138516eo.pdf
Kowalczyk, Shelly, Suzanne Randolph, Shereitte Stokes, and Stefanie Winston. 2014. "Evidence from the Field: Findings on Issues Related to Planning, Implementing and Evaluating Gender-Based Programs." *Evaluation and Program Planning*, 51, 35–44.
Kriger, Norma J. 1992. *Zimbabwe's Guerilla War: Peasant Voices*. Cambridge: Cambridge University Press.
Kriger, Norma J. 2003. *Guerrilla Veterans in Postwar Zimbabwe: Symbolic and Violent Politics, 1980–1987*. Cambridge: Cambridge University Press.
L'Engle, K., H. Vahdat, E. Ndakidemi, C. Lasway and T. Zan. 2013. "Evaluating Feasibility, Reach and Potential Impact of a Text Message Family Planning Information Service in Tanzania." *Contraception*, 87(2), 251–6.
Linde, D.S., M. Andersen, K. Mwaisselage, R. Manongi, S. Kajaer, and V. Rasch. 2017. "Text Messages to Increase Attendance to Follow-Up Cervical Cancer Screening Appointments among HPV-Positive Tanzanian Women (Connected2Care): Study Protocol for a Randomised Controlled Trial." *Trials*, 18, 555.
Lorde, Audre. 1984. "Age, Race, Class, and Sex: Women Defining Difference." In *Sister Outsider*, pp. 114–23. Trumansburg, NY: Crossing Press.

Lorde, Audre. 1993 [1984]. *Sister Outsider*. New York, NY: Quality Paperback Book Club.

Lund S., M. Hemed, B.B. Nielsen, A. Said, K. Said, M.H. Makungu, et al. 2012. "Mobile Phones as a Health Communication Tool to Improve Skilled Attendance at Delivery in Zanzibar: A Cluster-Randomised Controlled Trial." *BJOG: An International Journal of Obstetrics & Gynaecology*, 119(10), 1256–64.

Lwoga, E. and N. Lwoga. 2017. "User Acceptance of Mobile Payment: The Effects of User-centric Security, System Characteristics and Gender." *Electronic Journal of Information Systems in Developing Countries*, 81(1), 1–24.

Lyons, Tanya. 2004. *Guns and Guerilla Girls: Women in the Zimbabwean National Liberation Struggle*. Trenton, NJ: Africa World Press.

Madebwe, Crescentia and Victor Madebwe. 2011. "The Gender Dynamics of Zimbabwe's Fast Track Land Reform Programme." *The Dyke: A Journal of Midlands State University*, 5(1), 87–103.

Mafa, O., E.S. Gudhlanga, N. Manyeruke, E. Matavire, and J. Mpofu. 2015. *Gender Politics and Land Use in Zimbabwe: 1980–2012*. Dakar: CODESRIA.

Mama, Amina. 1995. *Beyond the Masks: Race, Gender and Subjectivity*. New York, NY: Routledge.

Mama, Amina. 1998. "Khaki in the Family: Gender Discourses and Militarism in Nigeria." *African Studies Review*, 41(2), 1–17.

Mama, Amina. 2002. "Editorial." *Feminist Africa: Intellectual Politics*, 1.

Masina, Lameck. 2010. "Push for More Women MPs Stymied in Malawi Voice of America News." *VOA*, December 28. Accessed at: www.voanews.com/a/push-women-mpsstymied-malawi/4182922.html

Matfess, H. 2017. "Three Years Later, A Look at the #BringBackOurGirls Catch-22." *The Daily Beast*. Accessed at: www.thedailybeast.com/three-years-later-a-look-at-thebringbackourgirls-catch-22

Mbilinyi, Marjorie. 1991. "Beyond Oppression and Crisis: A Gendered Analysis of Agrarian Structure and Change." CODESRIA Workshop on Gender Analysis and African Social Science, Dakar, Senegal, September 16–20.

McFadden, Patricia. 1992. "Nationalism and Gender Issues in South Africa." *Journal of Gender Studies*, 1, 510.

McFadden, Patricia. 1994. "Ethnicity in Africa and African Female Authenticity." *SAPEM*, 8(1), 33–4.

McFadden, Patricia. 2005. "Becoming Postcolonial: African Women Changing the Meaning of Citizenship." *Meridians: Feminism, Race, Transnationalism*, 6(1), 1–18.

McFadden, Patricia. 2007a. "War Through a Feminist Lens." *Ms*, 17(1), 14–15.

McFadden, Patricia. 2007b. "African Feminist Perspectives of PostColoniality." *Black Scholar*, 37, 36–42.

McFadden, Patricia. 2008. "Globalizing Resistance: Crafting and Strengthening African Feminist Solidarities." *Black Scholar*, 38(2/3), 19–20.

McFadden, Patricia. 2010. "Challenging Empowerment." *Development*, 53, 161–4.

McFadden, Patricia, and Keshia Nicole Abraham (eds). 1999. *Reflections on Gender Issues in Africa*. Harare: SAPES.

McFadden, Patricia, and Jenifer Chiriga (eds). 1998. *Gender in Southern Africa: A Gendered Perspective*. Harare: SAPES.

Meekers, Dominique, Ronan Van Rossem, Martha Silva, and Andrew Koleros. 2007. "The Reach and Effect of Radio Communication Campaigns on Condom Use in Malawi." *Studies in Family Planning*, 38(2), 113–20.

Ministry of Health (MoH) [Malawi] and ICF International. 2014. Malawi Service Provision Assessment (MSPA) 2013–2014. Lilongwe, Malawi, and Rockville, MD, USA: MoH and ICF International.

Ministry of Health and Social Welfare (MoHSW) [Tanzania Mainland], Ministry of Health (MoH) [Zanzibar], National Bureau of Statistics (NBS), Office of the Chief Government Statistician (OCGS), and ICF International 2015. Tanzania Service Provision Assessment Survey (TSPA) 2014–2015. Dar es Salaam, Tanzania, and Rockville, MD: MoHSW, MoH, NBS, OCGS, and ICF International.

Ministry of Health, Community Development, Gender, Elderly and Children (MoHCDGEC) [Tanzania Mainland], Ministry of Health (MoH) [Zanzibar], National Bureau of Statistics (NBS), Office of the Chief Government Statistician (OCGS), and ICF. 2016. Tanzania Demographic and Health Survey and Malaria Indicator Survey (TDHS-MIS) 2015–2016. Dar es Salaam, Tanzania, and Rockville, MD: MoHCDGEC, MoH, NBS, OCGS, and ICF.

Mlyakado, Budeba Petro. 2012. "Gender and Education Opportunities in Tanzania: Do We Bridge the Gap of Quality?" *Academic Research International*, 3(3), 246–55. Accessed at: www.savap.org.pk/journals/ARInt./Vol.3(3)/2012(3.3–29).pdf

Moyo, S. 1995. *The Land Question in Zimbabwe*. Harare: Sapes Books.

Moyo, S. 2000. *Land Reform Under Structural Adjustment in Zimbabwe*. Uppsala: The Nordic Africa Institute.

Moyo, S. 2007. "The Radicalized State: Zimbabwe's Interrupted Revolution." *Review of African Political Economy (ROAPE)*, 111, 103–21.

Moyo, S. 2008. *African Land Questions, Agrarian Transitions and the State: Contradictions of Neoliberal Land Reforms*. Dakar: CODESRIA Greenbook.

Moyo, S. and Kojo Sebastian Amanor (eds). 2008. *Land and Sustainable Development in Africa*. London: Zed Books.

Moyo, S. and Paris Yeros. 2001. *The Politics of Land Distribution and Race Relations in Southern Africa*. New York, NY: UNRISD.

Moyo, S., J. Makumbe, and B. Raftopoulos. 2000. *NGOs, the State & Politics in Zimbabwe*. Harare: SAPES Books.

Mtambalike, Maria J. 2013. "Domestic Violence Against Women in Tanzania, The NeverEnding Story." *We Write For Rights*. Accessed at: https://wewriteforrights.wordpress.com/2013/08/11/domestic-violence-against-women-intanzania-the-never-ending-story/

Murphy, Laura and A. Priebe 2011. "'My co-wife can borrow my mobile phone': Gendered Geographies of Cell Phone Usage and Significance for Rural Kenyans." *Gender, Technology and Development*, 15(1), 1–23.

Mutungi, Winnie. 2016. "Tanzania: Will the New Constitution Protect Women's Rights in Tanzania?" South African Institute of International Affairs. *All Africa*. Accessed at: http://allafrica.com/stories/201611160049.html

National Bureau of Statistics (NBS) [Tanzania] and ICF Macro. 2011. *Tanzania Demographic and Health Survey Country Report 2010*. Dar es Salaam, Tanzania: NBS and ICF Macro.

National Statistical Office (NSO) [Malawi] and ICF. 2017. Malawi Demographic and Health Survey 2015–2016. Zomba, Malawi, and Rockville, MD: NSO and ICF.

National Statistical Office (NSO) and ICF Macro. 2011. Malawi Demographic and Health Survey 2010. Zomba, Malawi, and Calverton, MD: NSO and ICF Macro.

National Statistical Office (NSO) [Malawi], and ORC Macro. 2005. Malawi Demographic and Health Survey 2004. Calverton, MD: NSO and ORC Macro.

Ndiweni, N. and J.M. Finch. 1995. "Effects of In Vitro Supplementation of Bovine Mammary Gland Macrophages and Peripheral Blood Lymphocytes with α-tocopherol and Sodium Selenite: Implications for Udder Defenses." *Veterinary Immunology and Immunopathology*, 47(1), 111–21.

Ndiweni, N. and J.M. Finch. 1996. "Effects of In Vitro Supplementation with α-tocopherol and Selenium on Bovine Neutrophil Functions: Implications for Resistance to Mastitis." *Veterinary Immunology and Immunopathology*, 51(1), 67–78.

Ntata, Pierson, Adamson Muula, Seter Siziya, and Edrinnie Kayambazinthu. 2008. "Health/HIV Knowledge and Preventive Behaviours among University Students." *SAHARA-J: Journal of Social Aspects of HIV/AIDS*, 5(4), 201–5.

Obbo, Christine. 1999. "Reviewed Work: *Reinventing Africa: Matriarchy, Religion and Culture* by Ifi Amadiume." *The Journal of Modern African Studies*, 37(3), 516–18.

Odhiambo, Agnes. 2014. "Promoting Gender Equality in Malawi." October 11. *Human Rights Watch Dispatches*. Accessed at: www.hrw.org/news/2014/10/11/dispatchespromoting-gender-equality-malawi

Oguibe, Olu. 2003. *The Culture Game*. Minneapolis: University of Minnesota Press.

Ogunyemi, Chikwenye. 1996. *Africa Wo/Man Palava: The Nigerian Novel by Women*. Chicago, IL: University of Chicago Press.

Oyeyemi, S.O. and R. Wynn. 2014. "Giving Cell Phones to Pregnant Women and Improving Services may Increase Primary Health Facility Utilization: A Case-Control Study of a Nigeria Project." *Reproductive Health*, 11(8), 1–8.

Pew Research Center. Cell Phones in Africa: Communication Lifeline. *Texting Most Common Activity, but Mobile Money Popular in Several Countries*. April 15, 2015. Accessed at: www.pewglobal.org/2015/04/15/cell-phones-in-africa-communication-lifeline/

Pew Research Center. *Spring 2014 Global Attitudes Survey*. Q7a-b.

Physicians for Human Rights. 2009. *Health in Ruins: A Man-Made Disaster in Zimbabwe*.

Raaber, Natalie 2007. "Media as an Advocacy Tool to Tackle Violence Against Women" *IWTC Women's Globalnet* 329.
Rabaka, Reiland. 2010. *Epistemic Apartheid: W.E.B. Du Bois and the Disciplinary Decadence of Sociology*. Lanham, MD: Lexington Books.
"Raising Women Participation in Politics: Issues to Consider." 2017. *Zimbabwe Independent*, September 22. Accessed at: www.theindependent.co.zw/2017/09/22/raisingwomen-participation-politics-issues-consider/
Risman, Barbara J. 2004. "Gender as a Social Structure: Theory Wrestling with Activism." *Gender & Society*, 18(4), 429–50.
Roy-Campbell, Zaline M. 2001. *Empowerment Through Language: The African Experience – Tanzania and Beyond*. Trenton, NJ: Africa World Press.
Rutashobya, Lettice Kinunda.1998. *Women Entrepreneurship in Tanzania: Entry and Performance Barriers*. Addis Ababa: OSSREA.
Sanya, B.N. and P.W. Odero. 2017. "Feminist Articulations, Social Literacies, and Ubiquitous Mobile Technology Use in Kenya." *Policy Futures in Education*, 15(3), 309–26.
Sassen, Saskia. 2003. "The Participation of States and Citizens in Global Governance" Symposium- Globalization and Governance: The Prospects for Democracy. *Journal of Global Legal Studies*, 10(5):1.
Sassen, Saskia. 2005. "The Global City: Introducing a Concept." *Brown Journal of World Affairs*, 9(2), 27–43.
Sassen, Saskia. 2017. "On How the Powerless Can 'Hack' Global Cities." *New Frontier*. August 10. Accessed at: https://archpaper.com/2017/08/saskia-sassen-hack-global-cities/
Seedat, M., A. Niekerk, R. Jewkes, R. Suffla, and K. Ratele . 2009. "Violence and Injuries in South Africa: Prioritising an Agenda for Prevention." *The Lancet*, 374(9694), 1011–22.
Semu, Linda. 2002. "Kamuzu's Mbumba: Malawi Women's Embeddedness to Culture in the Face of International Political Pressure and Internal Legal Change." *Africa Today*, 49(2), 77–99.
Semu, Linda L. 2005. *The Interplay of State, Family Structure and Land: A Study on Women and Children's Well-being in Matrilineal Households in Southern Malawi*. Dissertation.
Semu, Linda L. 2011. "Triangulation and Mixed Method Research Design in Practice: The Malawi Case Study on the Interplay of State Policy, Family Structure and Land on Women and Children's Well-Being." *The International Journal of Interdisciplinary Social Sciences*, 6(1), 57–71.
Skalli, L.H. 2014. "Young Women and Social Media Against Sexual Harassment in North Africa." *The Journal of North African Studies*, 19(2), 244–58.
Stark, L. 2013. "Transactional Sex and Mobile Phones in a Tanzanian Slum." *Suomen Antropologi*, 38(1), 12–20.
Tedre, M., V. Chuma, and M. Apiola. 2013. "Survey on Mobile Phone Battery End-Life in Tanzania." Presented at *AFRICON* 2013, pp. 1–5. Accessed at: https://ieeexplore.ieee.org/document/6757653/
Tolofari, Sowaribi. 2005. "New Public Management and Education." *Policy Futures in Education*, 3(1), 75–89.

Uchendu, Victor Chikezie. 1965. *The Igbo of Southeast Nigeria*. New York, NY: Holt, Rinehart and Winston.

Uchendu, V. 2007. "Ezi Na Ulo: The Extended Family in Igbo Civilization." *Dialectical Anthropology*, 31(1–3), 167–219.

UN DESA. 2016. "People-Centered Governance: Why and How Should We Increase Public Participation and Engagement." Compendium of Innovative Practices in Public Governance and Administration for Sustainable Development. New York, NY: United Nations. https://doi.org/10.18356/8945f0ef-en.

United Nations Educational, Scientific and Cultural Organization (UNESCO) and International Social Science Council. *World Social Science Report 2010: Knowledge Divides*. Paris: UNESCO.

United Nations Statistics Division. Deparment of Economic and Social Affairs. n.d.. "Millennium Development Goals Indicators: Mobile-Cellular Subscriptions." Accessed at: https://millenniumindicators.un.org/unsd/mdg/Metadata.aspx?IndicatorId=0&SeriesId=780

Usdin, S., Nicola Christofides, Lebo Malepe, and Aadielah Maker. 2000. "The Value of Advocacy in Promoting Social Change: Implementing the New Domestic Violence Act in South Africa." *Reproductive Health Matters*, 8(16), 55–66.

Uttal, Lynet and Gloria Holguin Cuádraz. 1999. "Intersectionality and In-Depth Interviews: Methodological Strategies for Analyzing Race, Class, and Gender." *Race, Gender and Class*, 6, 156–86.

Van Sertima, I. (ed.). 1989. *Great African Thinkers: CheikhAnta Diop*. New Bruswick, NJ.: Transaction Books.

Viriri, Advice, Excellent Chireshe, and Tompson Makahamadze. 2011. "'Domestic Violence Act and the Apartheid of Gender': A Critical Analysis of the Perceptions of Christian Women in Masvingo Province of Zimbabwe." *The Dyke: A Journal of the Midlands State University*, 5(1), 22–45.

Waite, Gloria M. 2000. "Traditional Medicine and the Quest for National Identity in Zimbabwe." *Zambezia*, 27, 235–68.

Wekwete, Naomi N. 2014. "Gender and Economic Empowerment in Africa: Evidence and Policy." *Journal of African Economies*, 23(1), i87–i127.

Wekwete, Naomi N. and Charity Manyeruke. 2012. "Strengthening Research Capacity in Africa: Gender, Sexuality and Politics with a Strategic Focus on the Lives of Young Women." In A. Mama, J. Bennett, and H. Chigudu (eds), *Feminist Africa 17 Researching Sexuality with Young Women: Southern Africa*, pp. 91–101. Capetown: African Gender Institute.

Wekwete, Naomi N., H. Sanhokwe, W. Murenjekwa, F. Takavarasha, and N. Madzingira. 2014. "The Association between Spousal Gender Based Violence and Women's Empowerment among Currently Married Women Aged 15–49 in Zimbabwe: Evidence from the 2010–2011 Zimbabwe Demographic and Health Survey." *African Population Studies*, 28(3), 1413–31.

West, Michael O. 2016. Personal Communication.

Wiza Kumwenda, Gregory Kunyenje, Jimmie Gama, Jacqueline Chinkonde, Francis Martinson, Irving Hoffman, Mina Hosseinipour, and Nora

Rosenberg. 2017. "Information Management in Malawi's Prevention of Mother-to-Child Transmission (PMTCT) Program: Health Workers' Perspectives." *Malawi Med J.*, 29(4), 306–10. doi: 10.4314/mmj.v29i4.5.
Women in Politics Support Unit. 2008. Women Legislators in the 7th Session of the Zimbabwean Parliament. Parliament of Zimbabwe. Accessed at: http://archive.kubatana.net/docs/women/wipsu_women_legislators_in_parliament_080 922.pdf
Women of Zimbabwe Arise. 2007. "Dreaming of a New Zimbabwe: The People's Charter." Accessed at: http://wozazimbabwe.org/?page_id=42
World Bank. 2017. *CPIA Criteria 2017*. Accessed at: http://pubdocs.worldbank.org/en/203511467141304327/CPIA-Criteria-2017v1.pdf
World Bank. 2017. "World Development Indicators." Accessed at: http://data.worldbank.org/datacatalog/world-development-indicators
WUA/UNICEF Partnership. 2015. Child Rights Workshop in Women's University in Africa. Accessed at: www.wua.ac.zw/index.php/component/content/article/266
Zerai, Assata. 2000a. "Models for Unity Between Scholarship and Grassroots Activism." *Critical Sociology*, 28, 201–16.
Zerai, Assata. 2000b. "Agents of Knowledge and Action: Selected Africana Scholars and Their Contributions to the Understanding of Race, Class and Gender Intersectionality." *Cultural Dynamics*, 12, 182–222.
Zerai, Assata. 2014. *Hypermasculinity, State Violence, & Family Well-Being: An Africana Feminist Analysis of Child Health in Zimbabwe*. Trenton, NJ: Africa World Press.
Zerai, Assata. 2016. Keynote. "In the Traditions of Professor Victor C. Uchendu and Professor Ifi Amadiume, A Proposal for Expanding African Transnational Feminist Praxis: Creating a Database of Women's Scholarship Concerning Women's Health in Zimbabwe." Symposium to commemorate the 50th anniversary of the release of Professor Victor Uchendu's The Igbo of Southeastern Nigeria. May 2, 2016. Virginia Polytechnic Institute and State University.
Zerai, Assata. 2018. Chapter 1: "The Lives of Women and Children in East Africa" in Assata Zerai and Brenda N. Sanya (eds), *Safe Water, Sanitation and Early Childhood Malnutrition in East Africa: An Africana Feminist Analysis of the lives of Women and Children in Kenya, Tanzania and Uganda.*
Zerai, Assata and Rae Banks. 2002. *Dehumanizing Discourse, Anti-Drug Law and Policy in America: A "Crack Mother's Nightmare."* Aldershot, England: Ashgate Publishing.
Zerai, Assata and Rebecca Morrow. 2018. "Gender as Social Structure and Its Potential Impact on Safe Water and Sanitation Technologies in East Africa: An African Feminist Analysis" in Assata Zerai and Brenda N. Sanya (eds), *Safe Water, Sanitation and Early Childhood Malnutrition in East Africa: An Africana Feminist Analysis of the lives of Women and Children in Kenya, Tanzania and Uganda.*
Zerai, Assata and Joanna Perez. 2018. "Paying Serious Attention to Women's Scholarship to Influence Policy in East Africa" in Assata Zerai and Brenda

N. Sanya (eds), *Safe Water, Sanitation and Early Childhood Malnutrition in East Africa: An Africana Feminist Analysis of the lives of Women and Children in Kenya, Tanzania and Uganda.*

Zerai, Assata and Brenda N. Sanya. 2018. "Introduction" in Assata Zerai and Brenda N. Sanya (eds), *Safe Water, Sanitation and Early Childhood Malnutrition in East Africa: An Africana Feminist Analysis of the lives of Women and Children in Kenya, Tanzania and Uganda.*

Zerai, Assata, Rebecca Morrow, and Courtney Cuthbertson. 2018. "Environmental Contamination and Early Childhood Morbidity in Kenya, Tanzania, and Uganda" in Assata Zerai and Brenda N. Sanya (eds), *Safe Water, Sanitation and Early Childhood Malnutrition in East Africa: An Africana Feminist Analysis of the lives of Women and Children in Kenya, Tanzania and Uganda.*

Zerai, Assata, Joanna Perez, and Chenyi Wang. 2016. "A Proposal for Expanding Endarkened Transnational Feminist Praxis: Creating a Database of Women's Scholarship and Activism to Promote Health in Zimbabwe." *Qualitative Inquiry*, 23(2), 107–18.

Zimbabwe Central Statistical Office and Institute for Resource Development/ Macro Systems, Inc. 1990. *Zimbabwe Demographic and Health Survey 1988.*

Zimbabwe Central Statistical Office and Institute for Resource Development/ Macro Systems, Inc. 2012. *Zimbabwe Demographic and Health Survey 2010–2011.*

Index

For Product Safety Concerns and Information please contact our EU representative GPSR@taylorandfrancis.com
Taylor & Francis Verlag GmbH, Kaufingerstraße 24, 80331 München, Germany

www.ingramcontent.com/pod-product-compliance
Ingram Content Group UK Ltd.
Pitfield, Milton Keynes, MK11 3LW, UK
UKHW021247190926
475569UK00002BA/38
9781138559363